Environmental Chemistry

Environmental Chemistry

Edited by
Jonathan Ayers

Larsen & Keller
www.larsen-keller.com

Environmental Chemistry
Edited by Jonathan Ayers
ISBN: 978-1-63549-691-8 (Hardback)

Larsen & Keller
Published by Larsen and Keller Education,
5 Penn Plaza,
19th Floor,
New York, NY 10001, USA

Cataloging-in-Publication Data

Environmental chemistry / edited by Jonathan Ayers.
p. cm.
Includes bibliographical references and index.
ISBN 978-1-63549-691-8
1. Environmental chemistry. 2. Green chemistry. I. Ayers, Jonathan.
TD193 .E58 2018
577.14--dc23

Printed and bound in China.

For more information regarding Larsen and Keller Education and its products, please visit the publisher's website www.larsen-keller.com

Table of Contents

Preface

Environmental chemistry refers to the natural chemical and biochemical processes occurring in the nature. It concerns with the evaluation and analysis of the effects, reactions, sources, fates, etc. of chemical products present in the air, water, and soil. The methods used in this field are quantitative chemical analysis, atomic and mass spectrometry, etc. The topics covered in this textbook offer the reader new insights in the field of environmental chemistry. It aims to serve as a resource guide for students and experts alike and contribute to the growth of the discipline.

To facilitate a deeper understanding of the contents of this book a short introduction of every chapter is written below:

Chapter 1- Atmosphere extends to 30 km above the Earth's surface. It is composed of nitrogen in huge quantity, oxygen, water vapour, and other gases. The section minutely discusses the concepts of atmospheric pressure, atmospheric circulation, pollution and its effects as well. This chapter will provide an integrated understanding of environmental chemistry.

Chapter 2- The presence of harmful substances in the Earth's atmosphere is called air pollution. The chapter delves into air pollution while keeping chemical pollutants in focus. Air pollutants can be categorised into particulate matter, odors, and gaseous pollutants. The section strategically encompasses and incorporates the major components and key concepts of air pollution, providing a complete understanding.

Chapter 3- When water bodies are contaminated with substances which are harmful for biotic or non-biotic components of the Earth, it is referred to as water pollution. This degradation is studied in the context of aquatic toxicology, which deals with effects of chemicals on aquatic bodies. This chapter studies in detail water pollution and aquatic toxicology.

Chapter 4- The major components of soil and its environmental issues are discussed in this section. There are various elements that lead to the formation of soil starting from climate, parent material, organisms, time and topography. Soil is composed of minerals, plant roots, organic materials, and pore spacing. The chapter also explores the concept of soil erosion and other environmental issues related to soil.

I would like to share the credit of this book with my editorial team who worked tirelessly on this book. I owe the completion of this book to the never-ending support of my family, who supported me throughout the project.

Editor

1

An Overview of Composition of Earth's Atmosphere

Atmosphere extends to 30 km above the Earth's surface. It is composed of nitrogen in huge quantity, oxygen, water vapour, and other gases. The section minutely discusses the concepts of atmospheric pressure, atmospheric circulation, pollution and its effects as well. This chapter will provide an integrated understanding of environmental chemistry.

Atmosphere of Earth

99% of atmospheric gases, including water vapor, extend only 30 kilometers (kms) above earth's surface.

Most of our weather, however, occurs within the first 10 to 15 kms.

Composition of Atmosphere

- Nitrogen - 78%
- Oxygen - 21%
- Water Vapor – 0 to 4%
- Carbon Dioxide - .037%
- Other gases make up the rest

Temperature Inversion

- An increase in air temperature with height often called simply an inversion.
- Radiosonde – an instrument that measures the vertical profile of air temperature in the atmosphere (sometimes exceeding 100,000 ft).

Atmospheric Layers

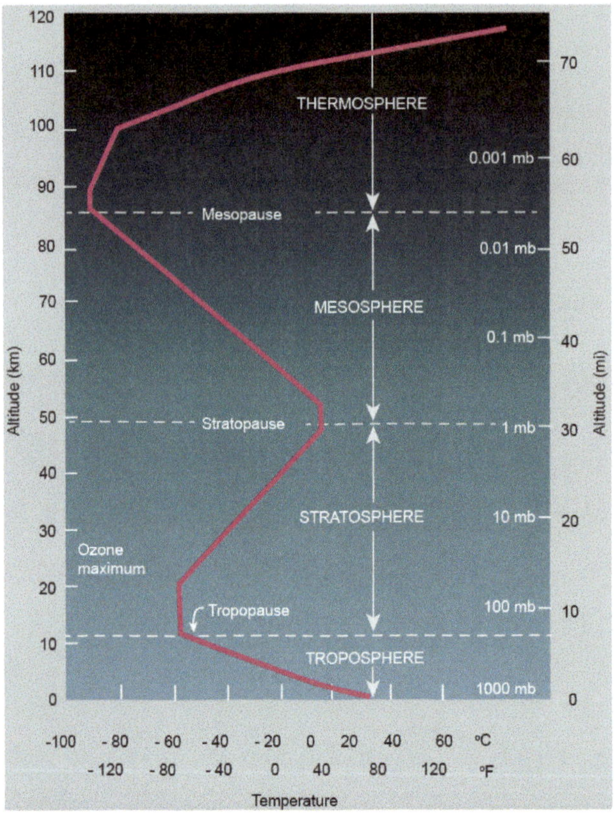

8 layers are defined by constant trends in average air temperature (which changes with pressure and radiation), where the outer exosphere is not shown.

1. Troposphere
2. Tropopause
3. Stratosphere
4. Stratopause
5. Mesosphere
6. Mesopause
7. Thermosphere
8. Exosphere

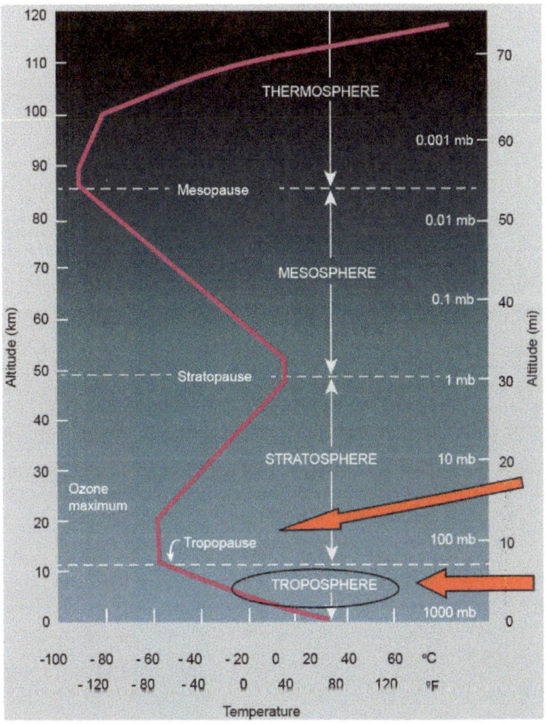

Tropopause separates Troposphere from Stratosphere. Generally higher in summer Lower in winter.

Troposphere – temperature decreases with height. Most of our weather occurs in this layer Varies in height around the globe, but Averages of about 11 km in height.

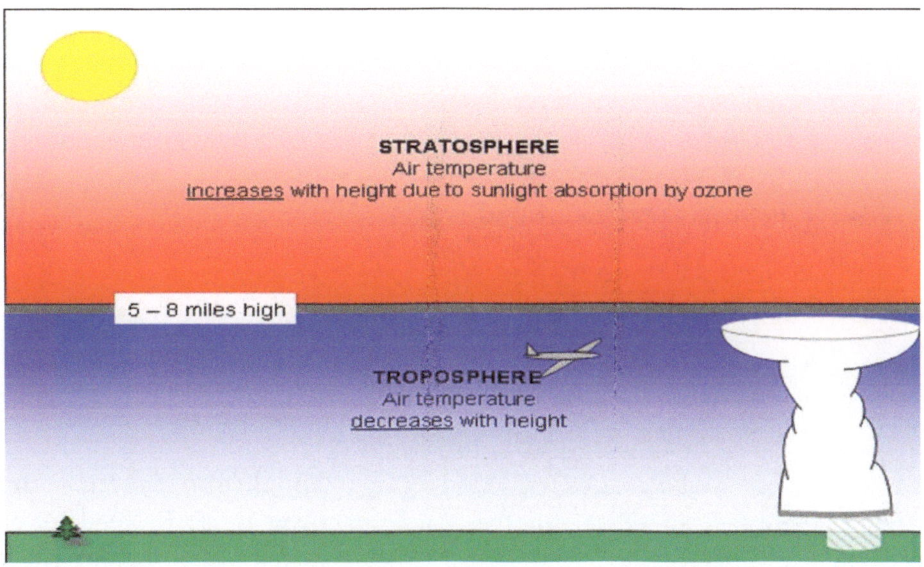

The troposphere is the lowest major atmospheric layer, and is located from the Earth's surface up to the bottom of the stratosphere . It has decreasing temperature with height (at an average rate of 3.5 °F per thousand feet (6.5°C per kilometer); whereas the stratosphere has either constant or slowly increasing temperature with height. The troposphere is where all of Earth's weather occurs. The boundary that divides the troposphere from the stratosphere is called the " tropopause ", located at an altitude of around 5 miles in the winter, to around 8 miles high in the summer, and as high as 11 or 12 miles in the deep tropics. When you see the top of a thunderstorm flatten out into an anvil cloud , like in the illustration above, it is usually because the updrafts in the storm are "bumping up against" the bottom of the stratosphere

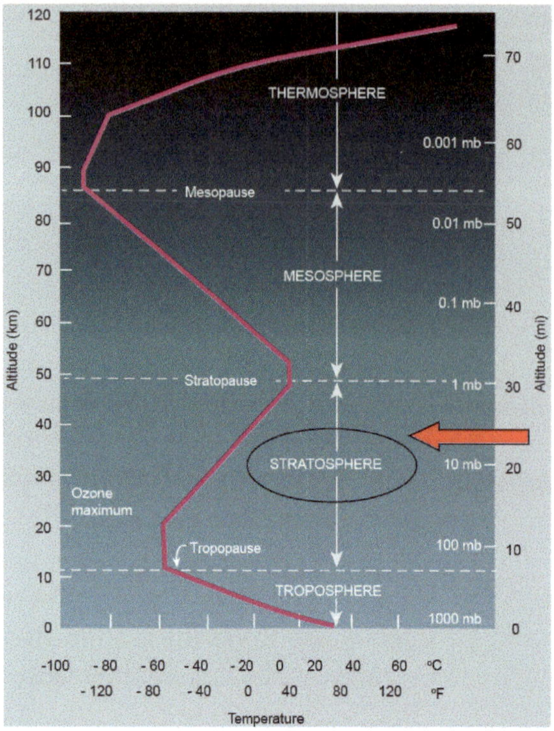

Stratosphere

Temperature inversion in stratosphere Ozone plays a major part in heating the air at this altitude

Temperature inversion in stratosphere Ozone plays a major part in heating the air at this altitude

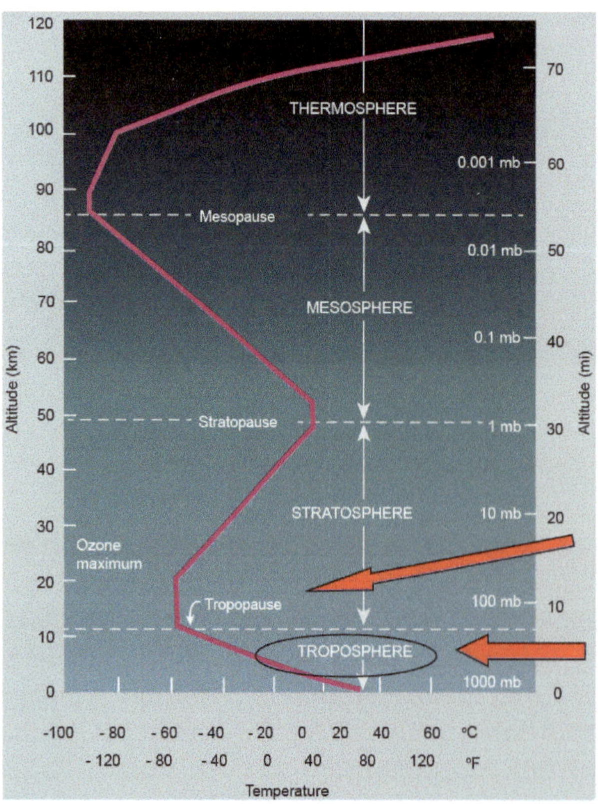

Mesosphere

Middle atmosphere – Air thin, pressure low, Need oxygen to live in this region. Air quite Cold -90°C (-130°F) near the top of mesosphere

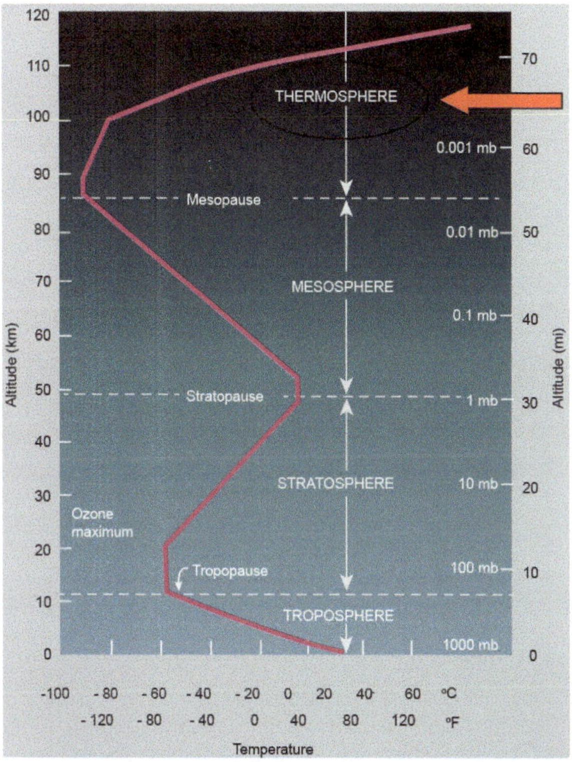

Thermosphere

"Hot layer" – oxygen molecules absorb energy from solar Rays warming the air. Very few atoms and molecules in this Region.

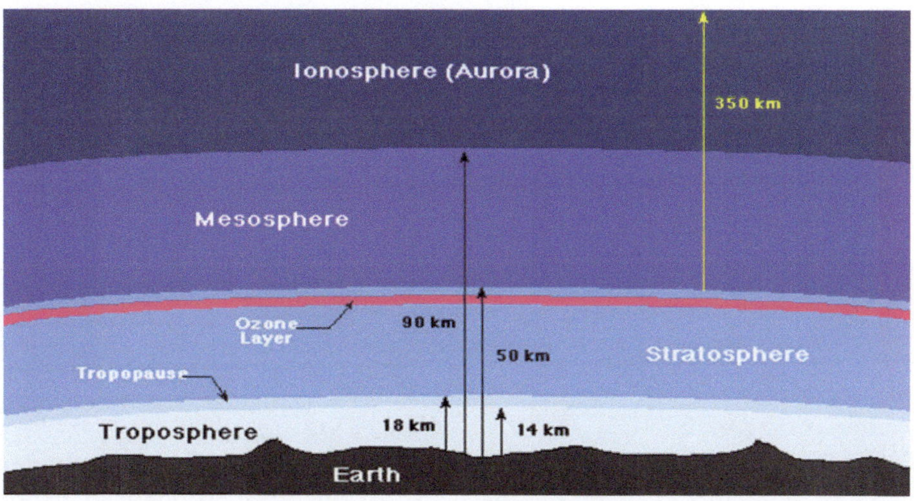

The Stratosphere and Ozone Layer

Above the troposphere is the stratosphere, where air flow is mostly horizontal. The thin ozone layer in the upper stratosphere has a high concentration of ozone, a particularly reactive form of oxygen. This layer is primarily responsible for absorbing the ultraviolet radiation from the Sun. The formation of this layer is a delicate matter, since as only when oxygen is produced in the atmosphere the ozone layer can form and prevent an intense flux of ultraviolet radiation from reaching the surface, where it is quite hazardous to the evolution of life. There is considerable recent concern that manmade flourocarbon compounds may be depleting the ozone layer, with dire future consequences for life on the Earth.

The Mesosphere and Ionosphere

Above the stratosphere is the mesosphere and above that is the ionosphere (or thermosphere), where many atoms are ionized (have gained or lost electrons so they have a net electrical charge). The ionosphere is very thin, but it is where aurora take place, and is also responsible for absorbing the most energetic photons from the Sun, and for reflecting radio waves, thereby making long-distance radio communication possible.

Atmospheric Mixture & Charge

Additional layers include:

a) The homosphere with 78% nitrogen and 21% oxygen

b) The poorly mixed heterosphere

c) The electrically charged ionosphere

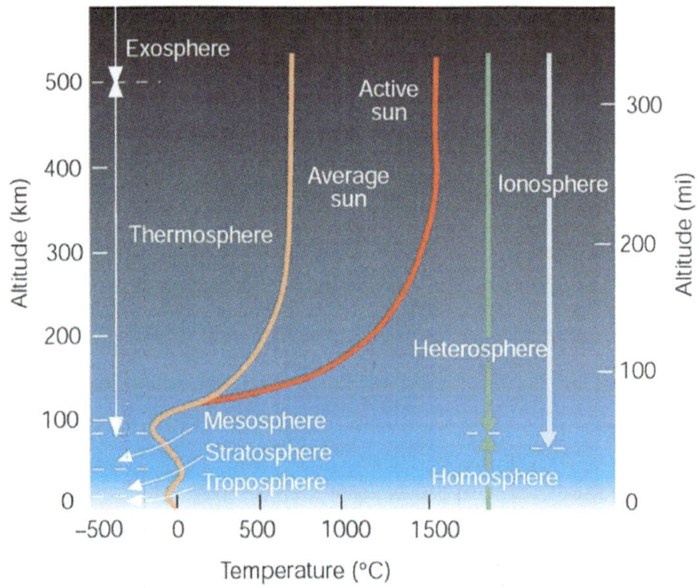

Atmospheric Gases

Ozone – is the primary ingredient of smog.

Nitrogen, oxygen, argon, water, vapor, carbon dioxide, and most other gases are invisible.

Clouds are not gas, but condensed vapor in the form of liquid droplets.

Ground based smog, which is visible, contains reactants of nitrogen and ozone.

Variable & Increasing Gases

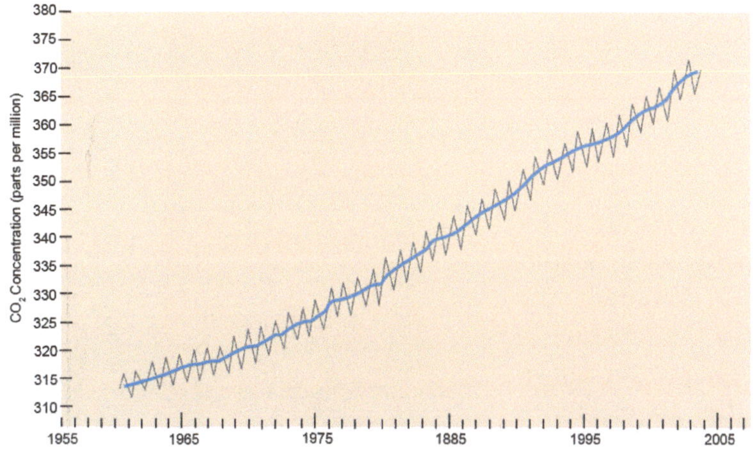

Nitrogen and oxygen concentrations experience little change, but carbon dioxide, methane, nitrous oxides, and chlorofluorocarbons are greenhouse gases experiencing discernable increases in concentration. CO_2 has risen more than 18% since 1958. Fossil fuels are the biggest problem.

Aerosols & Pollutants

Human and natural activities displace tiny soil, salt, and ash particles as suspended aerosols,

as well as sulfur and nitrogen oxides, and hydrocarbons as pollutants.

Pressure & Density

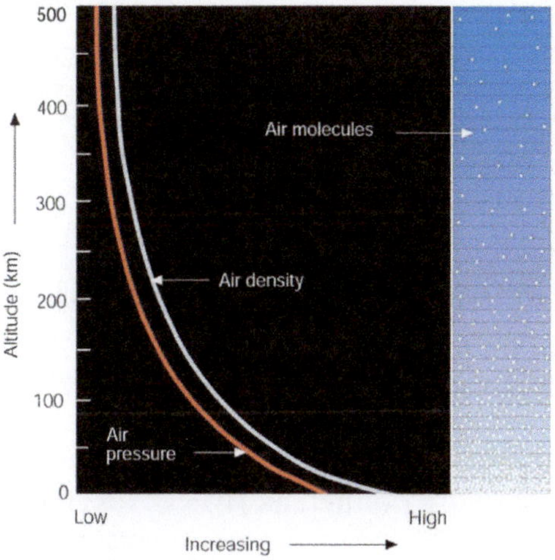

Gravity pulls gases toward earth's surface, and the whole column of gases weighs 14.7 psi at sea level, a pressure of 1013.25 mb or 29.92 in.Hg.

The amount of force exerted Over an area of surface is called Air pressure.

Air Density is The number of air Molecules in a given Space (volume)

Vertical Pressure Profile

Atmospheric pressure decreases rapidly with height. Climbing to an altitude of only 5.5 km where the pressure is 500 mb, would put you above one-half of the atmosphere's molecules.

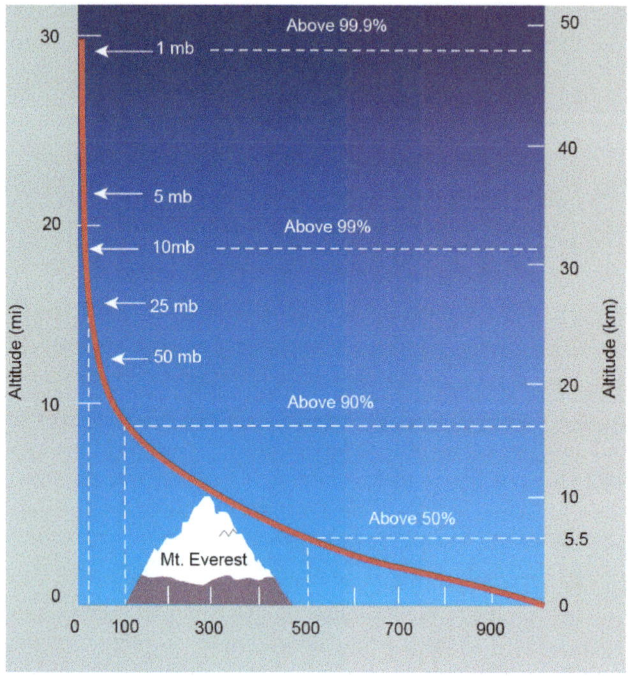

Atmospheric Pressure

Atmospheric pressure, sometimes also called barometric pressure, is the pressure exerted by the weight of air in the atmosphere of Earth (or that of another planet). In most circumstances atmospheric pressure is closely approximated by the hydrostatic pressure caused by the weight of air above the measurement point. As elevation increases, there is less overlying atmospheric mass, so that atmospheric pressure decreases with increasing elevation. On average, a column of air one square centimetre [cm^2] (0.16 sq in) in cross-section, measured from sea level to the top of the atmosphere, has a mass of about 1.03 kilograms (2.3 lb) and weight of about 10.1 newtons (2.3 lb_f). That force (across one square centimeter) is a pressure of 10.1 N/cm^2 or 101 kN/m^2. A column 1 square inch (6.5 cm^2) in cross-section would have a weight of about 14.7 lb (6.7 kg) or about 65.4 N.

Mechanism

Atmospheric pressure is caused by the gravitational attraction of the planet on the atmospheric gases above the surface, and is a function of the mass of the planet, the radius of the surface, and the amount of gas and its vertical distribution in the atmosphere. It is modified by the planetary rotation and local effects such as wind velocity, density variations due to temperature and variations in composition.

Standard Atmosphere

The standard atmosphere (symbol: atm) is a unit of pressure defined as 101325 Pa (1.01325 bar), equivalent to 760 mmHg (torr), 29.92 inHg and 14.696 psi.

Mean Sea Level Pressure

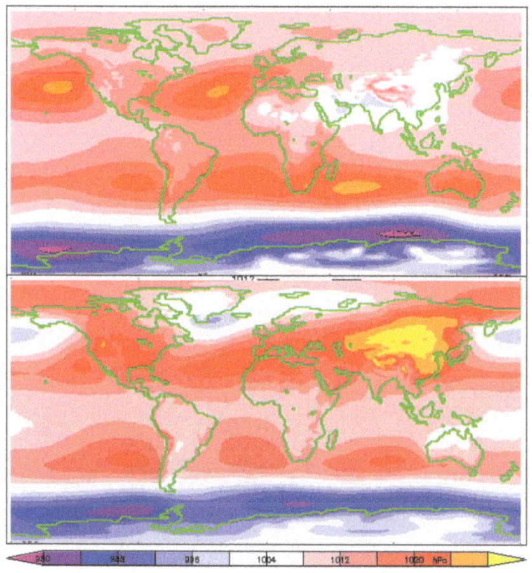

15 year average mean sea level pressure for June, July, and August (top) and December, January, and February (bottom). ERA-15 re-analysis.

Kollsman-type barometric aircraft altimeter (as used in North America) displaying an altitude of 80 ft (24 m).

The mean sea level pressure (MSLP) is the average atmospheric pressure at sea level. This is the atmospheric pressure normally given in weather reports on radio, television, and newspapers or on the Internet. When barometers in the home are set to match the local weather reports, they measure pressure adjusted to sea level, not the actual local atmospheric pressure.

The *altimeter setting* in aviation, is an atmospheric pressure adjustment.

Average *sea-level pressure* is 1013.25 mbar (101.325 kPa; 29.921 inHg; 760.00 mmHg). In aviation weather reports (METAR), QNH is transmitted around the world in millibars or hectopascals (1 hectopascal = 1 millibar), except in the United States, Canada, and Colombia where it is reported in inches (to two decimal places) of mercury. The United States and Canada also report *sea level pressure* SLP, which is adjusted to sea level by a different method, in the remarks section, not in the internationally transmitted part of the code, in hectopascals or millibars. However, in Canada's public weather reports, sea level pressure is instead reported in kilopascals.

In the US weather code remarks, three digits are all that are transmitted; decimal points and the one or two most significant digits are omitted: 1013.2 mbar (101.32 kPa) is transmitted as 132; 1000.0 mbar (100.00 kPa) is transmitted as 000; 998.7 mbar is transmitted as 987; etc. The highest *sea-level pressure* on Earth occurs in Siberia, where the Siberian High often attains a *sea-level pressure* above 1050 mbar (105 kPa; 31 inHg), with record highs close to 1085 mbar (108.5 kPa; 32.0 inHg). The lowest measurable *sea-level pressure* is found at the centers of tropical cyclones and tornadoes, with a record low of 870 mbar (87 kPa; 26 inHg).

Altitude Variation

A very local storm above Snæfellsjökull, showing clouds formed on the mountain by orographic lift

Variation in atmospheric pressure with altitude, computed for 15 °C and 0% relative humidity.

This plastic bottle was sealed at approximately 14,000 feet (4,300 m) altitude, and was crushed by the increase in atmospheric pressure —at 9,000 feet (2,700 m) and 1,000 feet (300 m)— as it was brought down towards sea level.

Pressure varies smoothly from the Earth's surface to the top of the mesosphere. Although the pressure changes with the weather, NASA has averaged the conditions for all parts of the earth year-round. As altitude increases, atmospheric pressure decreases. One can calculate the atmospheric pressure at a given altitude. Temperature and humidity also affect the atmospheric pressure, and it is necessary to know these to compute an accurate figure. The graph at right was developed for a temperature of 15 °C and a relative humidity of 0%.

At low altitudes above the sea level, the pressure decreases by about 1.2 kPa for every 100 meters. For higher altitudes within the troposphere, the following equation (the barometric formula) relates atmospheric pressure p to altitude h

$$p = p_0 \cdot \left(1 - \frac{L \cdot h}{T_0}\right)^{\frac{g \cdot M}{R_0 \cdot L}} \approx p_0 \cdot \left(1 - \frac{g \cdot h}{c_p \cdot T_0}\right)^{\frac{c_p \cdot M}{R_0}},$$

$$p \approx p_0 \cdot \exp\left(-\frac{g \cdot M \cdot h}{R_0 \cdot T_0}\right)$$

where the constant parameters are as described below:

Parameter	Description	Value
p_0	sea level standard atmospheric pressure	101325 Pa
L	temperature lapse rate, = g/c_p for dry air	0.0065 K/m
c_p	constant pressure specific heat	~ 1007 J/(kg•K)
T_0	sea level standard temperature	288.15 K
g	Earth-surface gravitational acceleration	9.80665 m/s²
M	molar mass of dry air	0.0289644 kg/mol
R_0	universal gas constant	8.31447 J/(mol•K)

Local Variation

Hurricane Wilma on 19 October 2005; 882 hPa (12.79 psi) in the storm's eye

Atmospheric pressure varies widely on Earth, and these changes are important in studying weather and climate.

Atmospheric pressure shows a diurnal or semidiurnal (twice-daily) cycle caused by global atmospheric tides. This effect is strongest in tropical zones, with an amplitude of a few millibars, and almost zero in polar areas. These variations have two superimposed cycles, a circadian (24 h) cycle and semi-circadian (12 h) cycle.

Records

The highest adjusted-to-sea level barometric pressure ever recorded on Earth (above 750 meters) was 1085.7 hPa (32.06 inHg) measured in Tosontsengel, Mongolia on 19 December 2001. The highest adjusted-to-sea level barometric pressure ever recorded (below 750 meters) was at Agata in Evenk Autonomous Okrug, Russia (66°53'N, 93°28'E, elevation: 261 m, 856 ft) on 31 December 1968 of 1083.8 hPa (32.005 inHg). The discrimination is due to the problematic assumptions (assuming a standard lapse rate) associated with reduction of sea level from high elevations.

The Dead Sea, the lowest place on Earth at 430 metres (1,410 ft) below sea level, has a correspondingly high typical atmospheric pressure of 1065 hPa.

The lowest non-tornadic atmospheric pressure ever measured was 870 hPa (0.858 atm; 25.69 inHg), set on 12 October 1979, during Typhoon Tip in the western Pacific Ocean. The measurement was based on an instrumental observation made from a reconnaissance aircraft.

Measurement Based on Depth of Water

One atmosphere (101 kPa or 14.7 psi) is the pressure caused by the weight of a column of fresh water of approximately 10.3 m (33.8 ft). Thus, a diver 10.3 m underwater experiences a pressure of about 2 atmospheres (1 atm of air plus 1 atm of water). Conversely, 10.3 m is the maximum height to which water can be raised using suction under standard atmospheric conditions.

Low pressures such as natural gas lines are sometimes specified in inches of water, typically written as *w.c.* (water column) or *w.g.* (inches water gauge). A typical gas-using residential appliance in the US is rated for a maximum of 14 w.c., which is approximately 35 hPa. Similar metric units with a wide variety of names and notation based on millimetres, centimetres or metres are now less commonly used.

Boiling Point of Water

Boiling water

Pure water boils at 100 °C (212 °F) at earth's standard atmospheric pressure. The boiling point is the temperature at which the vapor pressure is equal to the atmospheric pressure around the water. Because of this, the boiling point of water is lower at lower pressure and higher at higher pressure. Cooking at high elevations, therefore, requires adjustments to recipes. A rough approximation of elevation can be obtained by measuring the temperature at which water boils; in the mid-19th century, this method was used by explorers.

Measurement and Maps

An important application of the knowledge that atmospheric pressure varies directly with altitude was in determining the height of hills and mountains thanks to the availability of reliable pressure measurement devices. While in 1774 Maskelyne was confirming Newton's theory of gravitation at and on Schiehallion in Scotland (using plumb bob deviation to show the effect of "gravity") and accurately measure elevation, William Roy using barometric pressure was able to confirm his height determinations, the agreement being to within one meter (3.28 feet). This was then a useful tool for survey work and map making and long has continued to be useful. It was part of the "application of science" which gave practical people the insight that applied science could easily and relatively cheaply be "useful".

Atmospheric Circulation

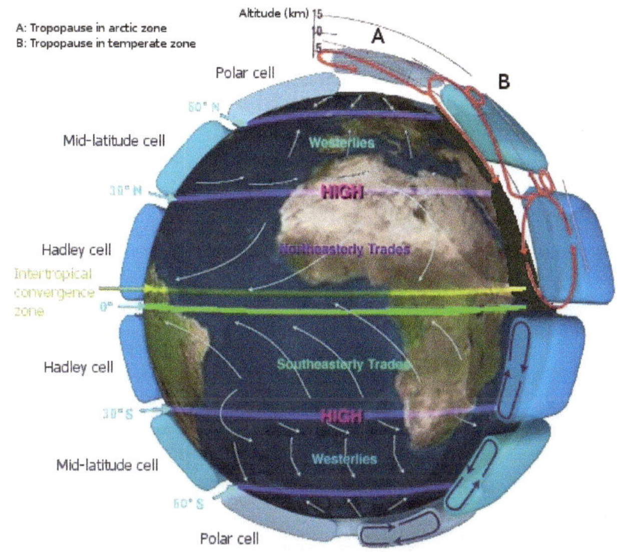

Idealised depiction (at equinox) of large scale atmospheric circulation on Earth.

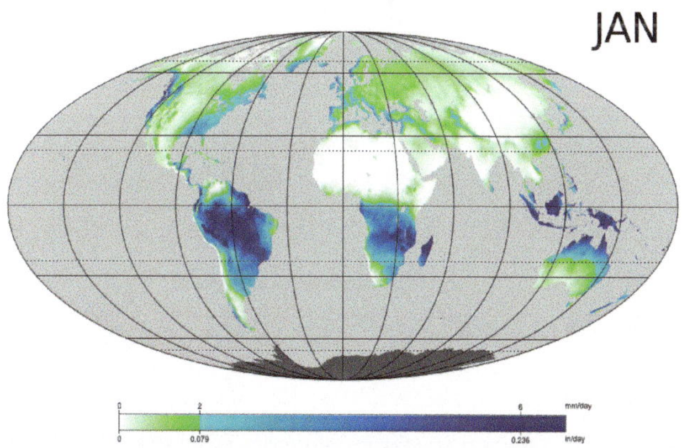

Long-term mean precipitation by month

Atmospheric circulation is the large-scale movement of air, and together with ocean circulation is the means by which thermal energy is redistributed on the surface of the Earth.

The Earth's atmospheric circulation varies from year to year, but the large scale structure of its circulation remains fairly constant. The smaller scale weather systems – mid-latitude depressions, or tropical convective cells – occur "randomly", and long range weather predictions of those cannot be made beyond ten days in practice, or a month in theory.

The Earth's weather is a consequence of its illumination by the Sun, and the laws of thermodynamics. The atmospheric circulation can be viewed as a heat engine driven by the Sun's energy, and whose energy sink, ultimately, is the blackness of space. The work produced by that engine causes the motion of the masses of air and in that process it redistributes the energy absorbed by the Earth's surface near the tropics to space and incidentally to the latitudes nearer the poles.

The large scale atmospheric circulation "cells" shift polewards in warmer periods (for example, interglacials compared to glacials), but remain largely constant as they are, fundamentally, a property of the Earth's size, rotation rate, heating and atmospheric depth, all of which change little. Over very long time periods (hundreds of millions of years), a tectonic uplift can significantly alter their major elements, such as the jet stream, and plate tectonics may shift ocean currents. During the extremely hot climates of the Mesozoic, a third desert belt may have existed at the Equator.

Latitudinal Circulation Features

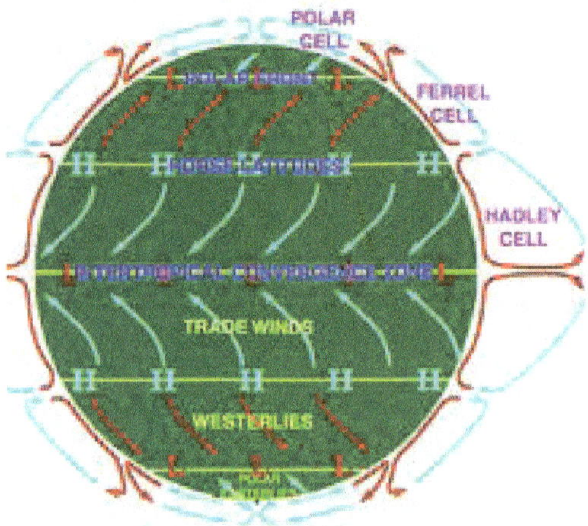

An idealised view of three large circulation cells showing surface winds

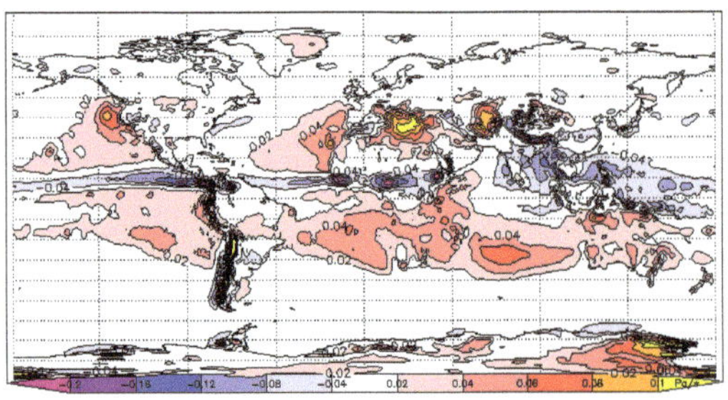

Vertical velocity at 500 hPa, July average. Ascent (negative values) is concentrated close to the solar equator; descent (positive values) is more diffuse but also occurs mainly in the Hadley cell.

The wind belts girdling the planet are organised into three cells in each hemisphere: the Hadley cell, the Ferrel cell, and the Polar cell. Those cells exist in both the northern and southern hemispheres. The vast bulk of the atmospheric motion occurs in the Hadley cell. The high pressure systems acting on the Earth's surface are balanced by the low pressure systems elsewhere. As a result, there is a balance of forces acting on the Earth's surface.

Hadley Cell

The ITCZ's band of clouds over the Eastern Pacific and the Americas as seen from space

The atmospheric circulation pattern that George Hadley described was an attempt to explain the trade winds. The Hadley cell is a closed circulation loop which begins at the equator. There, moist air is warmed by the Earth's surface, decreases in density and rises. A similar air mass rising on the other side of the equator forces those rising air masses to move poleward. The rising air creates a low pressure zone near the equator. As the air moves poleward, it cools, becomes more dense, and descends at about the 30th parallel, creating a high-pressure area. The descended air then travels toward the equator along the surface, replacing the air that rose from the equatorial zone, closing the loop of the Hadley cell. The poleward movement of the air in the upper part of the troposphere deviates toward the east, caused by the coriolis acceleration (a manifestation of conservation of energy). At the ground level however, the movement of the air toward the equator in the lower troposphere deviates toward the west, producing a wind from the east. The winds that flow to the west (from the east, easterly wind) at the ground level in the Hadley cell are called the Trade Winds.

Though the Hadley cell is described as located at the equator, in the northern hemisphere it shifts to higher latitudes in June and July and toward lower latitudes in December and January, which is the result of the Sun's heating of the surface. The zone where the greatest heating takes place is called the "thermal equator". As the southern hemisphere summer is December to March, the movement of the thermal equator to higher southern latitudes takes place then.

The Hadley system provides an example of a thermally direct circulation. The thermodynamic efficiency and power of the Hadley system, considered as a heat engine, is estimated at 200 terawatts.

Polar Cell

The Polar cell, likewise, is a simple system. Though cool and dry relative to equatorial air, the air masses at the 60th parallel are still sufficiently warm and moist to undergo convection and drive a thermal loop. At the 60th parallel, the air rises to the tropopause (about 8 km at this latitude) and moves poleward. As it does so, the upper level air mass deviates toward the east. When the air reaches the polar areas, it has cooled and is considerably denser than the underlying air. It descends, creating a cold, dry high-pressure area. At the polar surface level, the mass of air is driven toward the 60th parallel, replacing the air that rose there, and the polar circulation cell is complete. As the air at the surface moves toward the equator, it deviates toward the west. Again, the deviations of the air masses are the result of the Coriolis effect. The air flows at the surface are called the polar easterlies.

The outflow of air mass from the cell creates harmonic waves in the atmosphere known as Rossby waves. These ultra-long waves determine the path of the polar jet stream, which travels within the transitional zone between the tropopause and the Ferrel cell. By acting as a heat sink, the polar cell moves the abundant heat from the equator toward the polar regions.

The Hadley cell and the polar cell are similar in that they are thermally direct; in other words, they exist as a direct consequence of surface temperatures. Their thermal characteristics drive the weather in their domain. The sheer volume of energy that the Hadley cell transports, and the depth of the heat sink that is the polar cell, ensures that the effects of transient weather phenomena are not only not felt by the system as a whole, but — except under unusual circumstances — do not form. The endless chain of passing highs and lows which is part of everyday life for mid-latitude dwellers, at latitudes between 30 and 60° latitude, is unknown above the 60th and below the 30th parallels. There are some notable exceptions to this rule. In Europe, unstable weather extends to at least the 70th parallel north.

These atmospheric features are stable. Even though they may strengthen or weaken regionally over time, they do not vanish entirely.

The polar cell, orography and Katabatic winds in Antarctica, can create very cold conditions at the surface, for instance the coldest temperature recorded on Earth: −89.2 °C at Vostok Station in Antarctica, measured 1983.

Ferrel Cell

Part of the air rising at 60° latitude diverges at high altitude toward the poles and creates the polar cell. The rest moves toward the equator where it collides at 30° latitude with the high-level air of the Hadley cell. There it subsides and strengthens the high pressure ridges beneath. A large part of the energy that drives the Ferrel cell is provided by the polar and Hadley cells circulating on either side and that drag the Ferrel cell with it. The Ferrel cell, theorized by William Ferrel (1817–1891), is therefore a secondary circulation feature, whose existence depends upon the Hadley and polar cells on either side of it. It might be thought of as an eddy created by the Hadley and polar cells. The Ferrel cell is weak, and the air flow and temperatures within it are variable. For this reason, the mid-latitudes are sometimes known as the "zone of mixing." At high altitudes, the Ferrel cell overrides the Hadley and Polar cells. The air of the Ferrel cell that descends at 30° latitude returns poleward at the ground level, and as it does so it deviates toward the east. In the upper atmosphere of the Ferrel cell, the air moving toward the equator deviates toward the west. Both of those deviations, as in the case of the Hadley and polar cells, are driven by conservation of energy. As a result, just as the easterly Trade Winds are found below the Hadley cell, the Westerlies are found beneath the Ferrel cell. The forces driving the flow in the Ferrel cell are weak, and so the weather in that zone is variable. Thus, strong high-pressure areas which divert the prevailing westerlies, such as a Siberian high, can override the Ferrel cell, making it discontinuous.

While the Hadley and polar cells are truly closed loops, the Ferrel cell is not, and the telling point is in the Westerlies, which are more formally known as "the Prevailing Westerlies." The easterly Trade Winds and the polar easterlies have nothing over which to prevail, as their parent circulation cells are strong enough and face few obstacles either in the form of massive terrain features or high pressure zones. The weaker Westerlies of the Ferrel cell, however, can be disrupted. The local passage of a cold front may change that in a matter of minutes, and frequently does. As a result, at the surface, winds can vary abruptly in direction. But the winds above the surface, where they are less disrupted by terrain, are essentially westerly. A low pressure zone at 60° latitude that moves toward the equator, or a high pressure zone at 30° latitude that moves poleward, will accelerate the Westerlies of the Ferrel cell. A strong high, moving polewards may bring westerly winds for days.

The Ferrel cell is driven by the Hadley and Polar cells. It has neither a strong source of heat nor a strong sink to drive convection. As a result, the weather within the Ferrel cell is highly variable and is influenced by changes to the Hadley and Polar cells. The base of the Ferrel cell is characterized by the movement of air masses, and the location of those air masses is influenced in part by the location of the jet stream, even though it flows near the tropopause. Overall, the movement of surface air is from the 30th latitude to the 60th. However, the upper flow of the Ferrel cell is weak and not well defined.

In contrast to the Hadley and Polar systems, the Ferrel system provides an example of a thermally indirect circulation. The Ferrel system acts as a heat pump with a coefficient of performance of 12.1, consuming kinetic energy at an approximate rate of 275 terrawatts.

Longitudinal Circulation Features

While the Hadley, Ferrel, and polar cells (whose axes are oriented along parallels or latitudes) are the major features of global heat transport, they do not act alone. Temperature differences also drive a set of circulation cells, whose axes of circulation are longitudinally oriented. This atmospheric motion is known as zonal overturning circulation.

Latitudinal circulation is a result of the highest solar radiation per unit area (solar intensity) falling on the tropics. The solar intensity decreases as the latitude increases, reaching essentially zero at the poles. Longitudinal circulation, however, is a result of the heat capacity of water, its absorptivity, and its mixing. Water absorbs more heat than does the land, but its temperature does not rise as greatly as does the land. As a result, temperature variations on land are greater than on water. The Hadley, Ferrel, and polar cells operate at the largest scale of thousands of kilometers (synoptic scale). But, even at mesoscales (a horizontal range of 5 to several hundred kilometres), this effect is noticeable. During the day, air warmed by the relatively hotter land rises, and as it does so it draws a cool breeze from the sea that replaces the risen air. At night, the relatively warmer water and cooler land reverses the process, and a breeze from the land, of air cooled by the land, is carried offshore by night. This described effect is daily (diurnal).

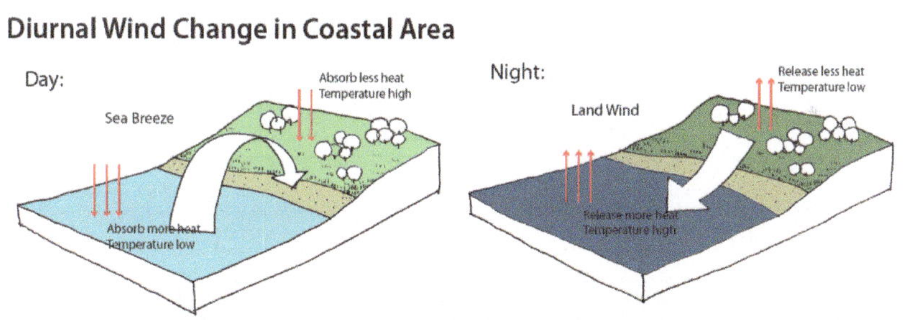

Diurnal wind change in local coastal area, also applies on the continental scale.

At the larger, synoptic, scale of oceans and continents, this effect is seasonal or even decadal. Warm air rises over the equatorial, continental, and western Pacific Ocean regions. When it reaches the tropopause, it cools and subsides in a region of relatively cooler water mass.

The Pacific Ocean cell plays a particularly important role in Earth's weather. This entirely ocean-based cell comes about as the result of a marked difference in the surface temperatures of the western and eastern Pacific. Under ordinary circumstances, the western Pacific waters are warm, and the eastern waters are cool. The process begins when strong convective activity over equatorial East Asia and subsiding cool air off South America's west coast creates a wind pattern which pushes Pacific water westward and piles it up in the western Pacific. (Water levels in the western Pacific are about 60 cm higher than in the eastern Pacific.)

Walker Circulation

The Pacific cell is of such importance that it has been named the Walker circulation after Sir Gilbert Walker, an early-20th-century director of British observatories in India, who sought a means of predicting when the monsoon winds of India would fail. While he was never successful in doing so, his work led him to the discovery of a link between the periodic pressure variations in the Indian Ocean, and those between the eastern and western Pacific, which he termed the "Southern Oscillation".

The movement of air in the Walker circulation affects the loops on either side. Under normal circumstances, the weather behaves as expected. But every few years, the winters become unusually warm or unusually cold, or the frequency of hurricanes increases or decreases, and the pattern sets in for an indeterminate period.

The Walker Cell plays a key role in this and in the El Niño phenomenon. If convective activity slows in the Western Pacific for some reason (this reason is not currently known), the climates of areas adjacent to the Western Pacific are affected. First, the upper-level westerly winds fail. This cuts off the source of returning, cool air that would normally subside at about 30° north latitude, and therefore the air returning as surface easterlies ceases. The consequence of this is twofold. Warm water ceases to surge into the eastern Pacific from the west (it was "piled" by past easterly winds) since there is no longer a surface wind to push it into the area of the west pacific. This and the corresponding effects of the Southern Oscillation result in long-term unseasonable temperatures and precipitation patterns in North and South America, Australia, and Southeast Africa, and the disruption of ocean currents.

Meanwhile, in the Atlantic, fast-blowing upper level Westerlies of the Hadley cell form, which would ordinarily be blocked by the Walker circulation and unable to reach such intensities. These winds disrupt the tops of nascent hurricanes and greatly diminish the number which are able to reach full strength.

El Niño – Southern Oscillation

El Niño and *La Niña* are opposite surface temperature anomalies of the Southern Pacific, which heavily influence the weather on a large scale. In the case of El Niño, warm surface water approaches the coasts of South America which results in blocking the upwelling of nutrient-rich deep water. This has serious impacts on the fish populations.

In the La Niña case, the convective cell over the western Pacific strengthens inordinately, resulting in colder than normal winters in North America and a more robust cyclone season in South-East Asia and Eastern Australia. There is also an increased upwelling of deep cold ocean waters and more intense uprising of surface air near South America, resulting in increasing numbers of drought occurrences, although fishermen reap benefits from the more nutrient-filled eastern Pacific waters.

Pollution

Pollution is the introduction of contaminants into the natural environment that cause adverse change. Pollution can take the form of chemical substances or energy, such as noise, heat or light. Pollutants, the components of pollution, can be either foreign substances/energies or naturally occurring contaminants. Pollution is often classed as point source or nonpoint source pollution.

Thermal oxidizers purify industrial air flows.

The litter problem on the coast of Guyana, 2010

History

Air pollution has always accompanied civilizations. Pollution started from prehistoric times when man created the first fires. According to an article in the journal *Science,* "soot" found on ceilings of prehistoric caves provides ample evidence of the high levels of pollution that was associated with inadequate ventilation of open fires." Metal forging appears to be a key turning point in the creation of significant air pollution levels outside the home. Core samples of glaciers in Greenland indicate increases in pollution associated with Greek, Roman and Chinese metal production, but at that time the pollution was comparatively small and could be handled by nature.

Urban Pollution

Air pollution in the US, 1973

The burning of coal and wood, and the presence of many horses in concentrated areas made the cities the cesspools of pollution. The Industrial Revolution brought an infusion of untreated chemicals and wastes into local streams that served as the water supply. King Edward I of England banned the burning of sea-coal by proclamation in London in 1272, after its smoke became a problem. But the fuel was so common in England that this earliest of names for it was acquired because it could be carted away from some shores by the wheelbarrow.

It was the industrial revolution that gave birth to environmental pollution as we know it today. London also recorded one of the earlier extreme cases of water quality problems with the Great Stink on the Thames of 1858, which led to construction of the London sewerage system soon afterward. Pollution issues escalated as population growth far exceeded view ability of neighborhoods to handle their waste problem. Reformers began to demand sewer systems, and clean water.

In 1870, the sanitary conditions in Berlin were among the worst in Europe. August Bebel recalled conditions before a modern sewer system was built in the late 1870s:

> "Waste-water from the houses collected in the gutters running alongside the curbs and emitted a truly fearsome smell. There were no public toilets in the streets or squares. Visitors, especially women, often became desperate when nature called. In the public buildings the sanitary facilities were unbelievably primitive. As a metropolis, Berlin did not emerge from a state of barbarism into civilization until after 1870."

The primitive conditions were intolerable for a world national capital, and the Imperial German government brought in its scientists, engineers and urban planners to not only solve the deficiencies but to forge Berlin as the world's model city. A British expert in 1906 concluded that Berlin represented "the most complete application of science, order and method of public life," adding "it is a marvel of civic administration, the most modern and most perfectly organized city that there is."

The emergence of great factories and consumption of immense quantities of coal gave rise to unprecedented air pollution and the large volume of industrial chemical discharges added to the growing load of untreated human waste. Chicago and Cincinnati were the first two American cities to enact laws ensuring cleaner air in 1881. Pollution became a major issue in the United States in the early twentieth century, as progressive reformers took issue with air pollution caused by coal burning, water pollution caused by bad sanitation, and street pollution caused by the 3 million horses who worked in American cities in 1900, generating large quantities of urine and manure. As historian Martin Melosi notes, The generation that first saw automobiles replacing the horses saw cars as "miracles of cleanliness.". By the 1940s, however, automobile-caused smog was a major issue in Los Angeles.

Other cities followed around the country until early in the 20th century, when the short lived Office of Air Pollution was created under the Department of the Interior. Extreme smog events were experienced by the cities of Los Angeles and Donora, Pennsylvania in the late 1940s, serving as another public reminder. Air pollution would continue to be a problem in England, especially later during the industrial revolution, and extending into the recent past with the Great Smog of 1952.

Awareness of atmospheric pollution spread widely after World War II, with fears triggered by reports of radioactive fallout from atomic warfare and testing. Then a non-nuclear event, The Great Smog of 1952 in London, killed at least 4000 people. This prompted some of the first major modern environmental legislation, The Clean Air Act of 1956.

Pollution began to draw major public attention in the United States between the mid-1950s and early 1970s, when Congress passed the Noise Control Act, the Clean Air Act, the Clean Water Act and the National Environmental Policy Act.

Smog Pollution in Taiwan

Severe incidents of pollution helped increase consciousness. PCB dumping in the Hudson River resulted in a ban by the EPA on consumption of its fish in 1974. Long-term dioxin contamination at Love Canal starting in 1947 became a national news story in 1978 and led to the Superfund legislation of 1980. The pollution of industrial land gave rise to the name brownfield, a term now common in city planning.

The development of nuclear science introduced radioactive contamination, which can remain lethally radioactive for hundreds of thousands of years. Lake Karachay, named by the Worldwatch Institute as the "most polluted spot" on earth, served as a disposal site for the Soviet Union throughout the 1950s and 1960s. Chelyabinsk, Russia, is considered the "Most polluted place on the planet".

Nuclear weapons continued to be tested in the Cold War, especially in the earlier stages of their development. The toll on the worst-affected populations and the growth since then in understanding about the critical threat to human health posed by radioactivity has also been a prohibitive complication associated with nuclear power. Though extreme care is practiced in that industry, the potential for disaster suggested by incidents such as those at Three Mile Island and Chernobyl pose a lingering specter of public mistrust. Worldwide publicity has been intense on those disasters. Widespread support for test ban treaties has ended almost all nuclear testing in the atmosphere.

International catastrophes such as the wreck of the Amoco Cadiz oil tanker off the coast of Brittany in 1978 and the Bhopal disaster in 1984 have demonstrated the universality of such events and the scale on which efforts to address them needed to engage. The borderless nature of atmosphere and oceans inevitably resulted in the implication of pollution on a planetary level with the issue of global warming. Most recently the term persistent organic pollutant (POP) has come to describe a group of chemicals such as PBDEs and PFCs among others. Though their effects remain somewhat less well understood owing to a lack of experimental data, they have been detected in various ecological habitats far removed from industrial activity such as the Arctic, demonstrating diffusion and bioaccumulation after only a relatively brief period of widespread use.

A much more recently discovered problem is the Great Pacific Garbage Patch, a huge concentration of plastics, chemical sludge and other debris which has been collected into a large area of the Pacific Ocean by the North Pacific Gyre. This is a less well known pollution problem than the others described above, but nonetheless has multiple and serious consequences such as increasing wildlife mortality, the spread of invasive species and human ingestion of toxic chemicals. Organizations such as 5 Gyres have researched the pollution and, along with artists like Marina DeBris, are working toward publicizing the issue.

Pollution introduced by light at night is becoming a global problem, more severe in urban centres, but nonetheless contaminating also large territories, far away from towns.

Growing evidence of local and global pollution and an increasingly informed public over time have given rise to environmentalism and the environmental movement, which generally seek to limit human impact on the environment.

Forms of Pollution

The Lachine Canal in Montreal, Quebec, Canada.

Blue drain and yellow fish symbol used by the UK Environment Agency to raise awareness of the ecological impacts of contaminating surface drainage.

The major forms of pollution are listed below along with the particular contaminant relevant to each of them:

- Air pollution: the release of chemicals and particulates into the atmosphere. Common gaseous pollutants include carbon monoxide, sulfur dioxide, chlorofluorocarbons (CFCs) and nitrogen oxides produced by industry and motor vehicles. Photochemical ozone and smog are created as nitrogen oxides and hydrocarbons react to sunlight. Particulate matter, or fine dust is characterized by their micrometre size PM_{10} to $PM_{2.5}$.
- Light pollution: includes light trespass, over-illumination and astronomical interference.
- Littering: the criminal throwing of inappropriate man-made objects, unremoved, onto public and private properties.
- Noise pollution: which encompasses roadway noise, aircraft noise, industrial noise as well as high-intensity sonar.
- Soil contamination occurs when chemicals are released by spill or underground leakage. Among the most significant soil contaminants are hydrocarbons, heavy metals, MTBE, herbicides, pesticides and chlorinated hydrocarbons.
- Radioactive contamination, resulting from 20th century activities in atomic physics, such as nuclear power generation and nuclear weapons research, manufacture and deployment.
- Thermal pollution, is a temperature change in natural water bodies caused by human influence, such as use of water as coolant in a power plant.
- Visual pollution, which can refer to the presence of overhead power lines, motorway billboards, scarred landforms (as from strip mining), open storage of trash, municipal solid waste or space debris.
- Water pollution, by the discharge of wastewater from commercial and industrial waste (intentionally or through spills) into surface waters; discharges of untreated domestic sewage, and chemical contaminants, such as chlorine, from treated sewage; release of waste and contaminants into surface runoff flowing to surface waters (including urban runoff and agricultural runoff, which may contain chemical fertilizers and pesticides); waste disposal and leaching into groundwater; eutrophication and littering.
- Plastic pollution: involves the accumulation of plastic products in the environment that adversely affects wildlife, wildlife habitat, or humans.

Pollutants

A pollutant is a waste material that pollutes air, water or soil. Three factors determine the severity of a pollutant: its chemical nature, the concentration and the persistence.

Cost of Pollution

Pollution has cost. Manufacturing activities that cause air pollution impose health and clean-up costs on the whole society, whereas the neighbors of an individual who chooses to fire-proof his home may benefit from a reduced risk of a fire spreading to their own houses. If external costs exist, such as pollution, the producer may choose to produce more of the product than would be produced if the producer were required to pay all associated environmental costs. Because responsibility or consequence for self-directed action lies partly outside the self, an element of externalization is involved. If there are external benefits, such as in public safety, less of the good may be produced than would be the case if the producer were to receive payment for the external benefits to others.

Sources and Causes

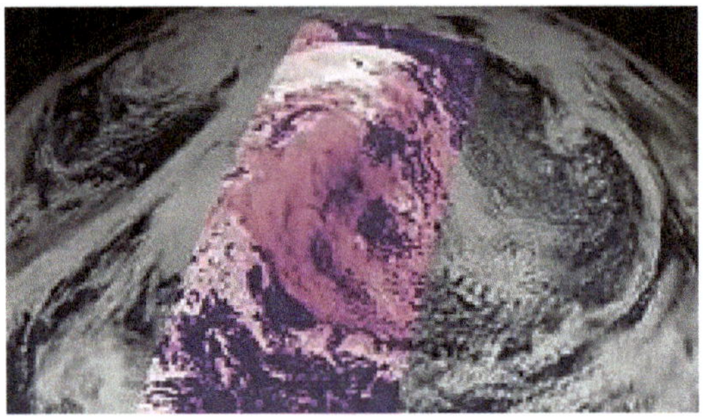

Air pollution produced by ships may alter clouds, affecting global temperatures.

Air pollution comes from both natural and human-made (anthropogenic) sources. However, globally human-made pollutants from combustion, construction, mining, agriculture and warfare are increasingly significant in the air pollution equation.

Motor vehicle emissions are one of the leading causes of air pollution. China, United States, Russia, India Mexico, and Japan are the world leaders in air pollution emissions. Principal stationary pollution sources include chemical plants, coal-fired power plants, oil refineries, petrochemical plants, nuclear waste disposal activity, incinerators, large livestock farms (dairy cows, pigs, poultry, etc.), PVC factories, metals production factories, plastics factories, and other heavy industry. Agricultural air pollution comes from contemporary practices which include clear felling and burning of natural vegetation as well as spraying of pesticides and herbicides

About 400 million metric tons of hazardous wastes are generated each year. The United States alone produces about 250 million metric tons. Americans constitute less than 5% of the world's population, but produce roughly 25% of the world's CO_2, and generate approximately 30% of world's waste. In 2007, China has overtaken the United States as the world's biggest producer of CO_2, while still far behind based on per capita pollution - ranked 78th among the world's nations.

An industrial area, with a power plant, south of Yangzhou's downtown, China

In February 2007, a report by the Intergovernmental Panel on Climate Change (IPCC), representing the work of 2,500 scientists, economists, and policymakers from more than 120 countries, said that humans have been the primary cause of global warming since 1950. Humans have ways to cut greenhouse gas emissions and avoid the consequences of global warming, a major climate report concluded. But to change the climate, the transition from fossil fuels like coal and oil needs to occur within decades, according to the final report this year from the UN's Intergovernmental Panel on Climate Change (IPCC).

Some of the more common soil contaminants are chlorinated hydrocarbons (CFH), heavy metals (such as chromium, cadmium–found in rechargeable batteries, and lead–found in lead paint, aviation fuel and still in some countries, gasoline), MTBE, zinc, arsenic and benzene. In 2001 a series of press reports culminating in a book called *Fateful Harvest* unveiled a widespread practice of recycling industrial byproducts into fertilizer, resulting in the contamination of the soil with various metals. Ordinary municipal landfills are the source of many chemical substances entering the soil environment (and often groundwater), emanating from the wide variety of refuse accepted, especially substances illegally discarded there, or from pre-1970 landfills that may have been subject to little control in the U.S. or EU. There have also been some unusual releases of polychlorinated dibenzodioxins, commonly called *dioxins* for simplicity, such as TCDD.

Pollution can also be the consequence of a natural disaster. For example, hurricanes often involve water contamination from sewage, and petrochemical spills from ruptured boats or automobiles. Larger scale and environmental damage is not uncommon when coastal oil rigs or refineries are involved. Some sources of pollution, such as nuclear power plants or oil tankers, can produce widespread and potentially hazardous releases when accidents occur.

In the case of noise pollution the dominant source class is the motor vehicle, producing about ninety percent of all unwanted noise worldwide.

Effects

Human Health

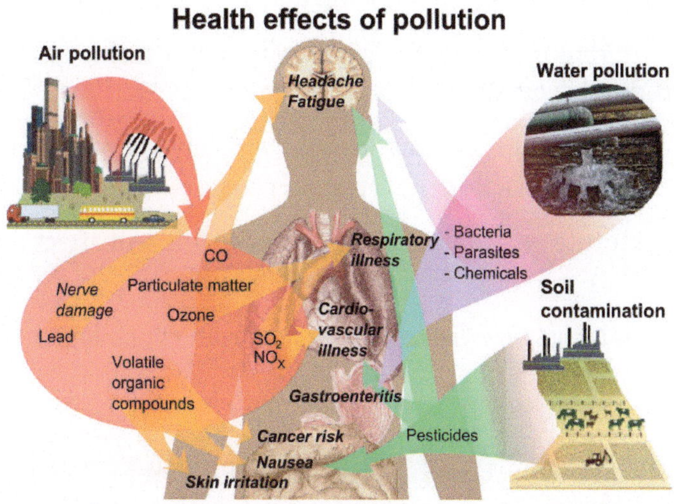

Overview of main health effects on humans from some common types of pollution.

Adverse air quality can kill many organisms including humans. Ozone pollution can cause respiratory disease, cardiovascular disease, throat inflammation, chest pain, and congestion. Water pollution causes approximately 14,000 deaths per day, mostly due to contamination of drinking water by untreated sewage in developing countries. An estimated 500 million Indians have no access to a proper toilet, Over ten million people in India fell ill with waterborne illnesses in 2013, and 1,535 people died, most of them children. Nearly 500 million Chinese lack access to safe drinking water. A 2010 analysis estimated that 1.2 million people died prematurely each year in China because of air pollution. The WHO estimated in 2007 that air pollution causes half a million deaths per year in India. Studies have estimated that the number of people killed annually in the United States could be over 50,000.

Oil spills can cause skin irritations and rashes. Noise pollution induces hearing loss, high blood pressure, stress, and sleep disturbance. Mercury has been linked to developmental deficits in children and neurologic symptoms. Older people are majorly exposed to diseases induced by air pollution. Those with heart or lung disorders are at additional risk. Children and infants are also at serious risk. Lead and other heavy metals have been shown to cause neurological problems. Chemical and radioactive substances can cause cancer and as well as birth defects.

Environment

Pollution has been found to be present widely in the environment. There are a number of effects of this:

- Biomagnification describes situations where toxins (such as heavy metals) may pass through trophic levels, becoming exponentially more concentrated in the process.
- Carbon dioxide emissions cause ocean acidification, the ongoing decrease in the pH of the Earth's oceans as CO_2 becomes dissolved.
- The emission of greenhouse gases leads to global warming which affects ecosystems in many ways.
- Invasive species can out compete native species and reduce biodiversity. Invasive plants can contribute debris and biomolecules (allelopathy) that can alter soil and chemical compositions of an environment, often reducing native species competitiveness.
- Nitrogen oxides are removed from the air by rain and fertilise land which can change the species composition of ecosystems.
- Smog and haze can reduce the amount of sunlight received by plants to carry out photosynthesis and leads to the production of tropospheric ozone which damages plants.
- Soil can become infertile and unsuitable for plants. This will affect other organisms in the food web.
- Sulfur dioxide and nitrogen oxides can cause acid rain which lowers the pH value of soil.

Environmental Health Information

The Toxicology and Environmental Health Information Program (TEHIP) at the United States National Library of Medicine (NLM) maintains a comprehensive toxicology and environmental health web site that includes access to resources produced by TEHIP and by other government agencies and organizations. This web site includes links to databases, bibliographies, tutorials, and other scientific and consumer-oriented resources. TEHIP also is responsible for the Toxicology Data Network (TOXNET) an integrated system of toxicology and environmental health databases that are available free of charge on the web.

TOXMAP is a Geographic Information System (GIS) that is part of TOXNET. TOXMAP uses maps of the United States to help users visually explore data from the United States Environmental Protection Agency's (EPA) Toxics Release Inventory and Superfund Basic Research Programs.

Worker Productivity

A number of studies show that pollution has an adverse effect on the productivity of both indoor and outdoor workers.

Regulation and Monitoring

To protect the environment from the adverse effects of pollution, many nations worldwide have enacted legislation to regulate various types of pollution as well as to mitigate the adverse effects of pollution.

Pollution Control

A Mobile Pollution Check Vehicle in India.

A dust collector in Pristina, Kosovo

Gas nozzle with vapor recovery

Pollution control is a term used in environmental management. It means the control of emissions and effluents into air, water or soil. Without pollution control, the waste products from overconsumption, heating, agriculture, mining, manufacturing, transportation and other human activities, whether they accumulate or disperse, will degrade the environment. In the hierarchy of controls, pollution prevention and waste minimization are more desirable than pollution control. In the field of land development, low impact development is a similar technique for the prevention of urban runoff.

Practices

- Recycling
- Reusing

- Waste minimisation
- Mitigating
- Preventing
- Compost

Pollution Control Devices

- Air pollution control
    - Thermal oxidizer
- Dust collection systems
    - Baghouses
    - Cyclones
    - Electrostatic precipitators
- Scrubbers
    - Baffle spray scrubber
    - Cyclonic spray scrubber
    - Ejector venturi scrubber
    - Mechanically aided scrubber
    - Spray tower
    - Wet scrubber
- Sewage treatment
    - Sedimentation (Primary treatment)
    - Activated sludge biotreaters (Secondary treatment; also used for industrial wastewater)
    - Aerated lagoons
    - Constructed wetlands (also used for urban runoff)
- Industrial wastewater treatment
    - API oil-water separators
    - Biofilters
    - Dissolved air flotation (DAF)

  - o Powdered activated carbon treatment
  - o Ultrafiltration
- Vapor recovery systems
- Phytoremediation

Perspectives

The earliest precursor of pollution generated by life forms would have been a natural function of their existence. The attendant consequences on viability and population levels fell within the sphere of natural selection. These would have included the demise of a population locally or ultimately, species extinction. Processes that were untenable would have resulted in a new balance brought about by changes and adaptations. At the extremes, for any form of life, consideration of pollution is superseded by that of survival.

For humankind, the factor of technology is a distinguishing and critical consideration, both as an enabler and an additional source of byproducts. Short of survival, human concerns include the range from quality of life to health hazards. Since science holds experimental demonstration to be definitive, modern treatment of toxicity or environmental harm involves defining a level at which an effect is observable. Common examples of fields where practical measurement is crucial include automobile emissions control, industrial exposure (e.g. Occupational Safety and Health Administration (OSHA) PELs), toxicology (e.g. LD_{50}), and medicine (e.g. medication and radiation doses).

"The solution to pollution is dilution", is a dictum which summarizes a traditional approach to pollution management whereby sufficiently diluted pollution is not harmful. It is well-suited to some other modern, locally scoped applications such as laboratory safety procedure and hazardous material release emergency management. But it assumes that the dilutant is in virtually unlimited supply for the application or that resulting dilutions are acceptable in all cases.

Such simple treatment for environmental pollution on a wider scale might have had greater merit in earlier centuries when physical survival was often the highest imperative, human population and densities were lower, technologies were simpler and their byproducts more benign. But these are often no longer the case. Furthermore, advances have enabled measurement of concentrations not possible before. The use of statistical methods in evaluating outcomes has given currency to the principle of probable harm in cases where assessment is warranted but resorting to deterministic models is impractical or infeasible. In addition, consideration of the environment beyond direct impact on human beings has gained prominence.

Yet in the absence of a superseding principle, this older approach predominates practices throughout the world. It is the basis by which to gauge concentrations of effluent for legal release, exceeding which penalties are assessed or restrictions applied. One such superseding principle is contained in modern hazardous waste laws in developed countries, as the process of diluting hazardous waste to make it non-hazardous is usually a regulated treatment process. Migration from pollution dilution to elimination in many cases can be confronted by challenging economical and technological barriers.

Greenhouse Gases and Global Warming

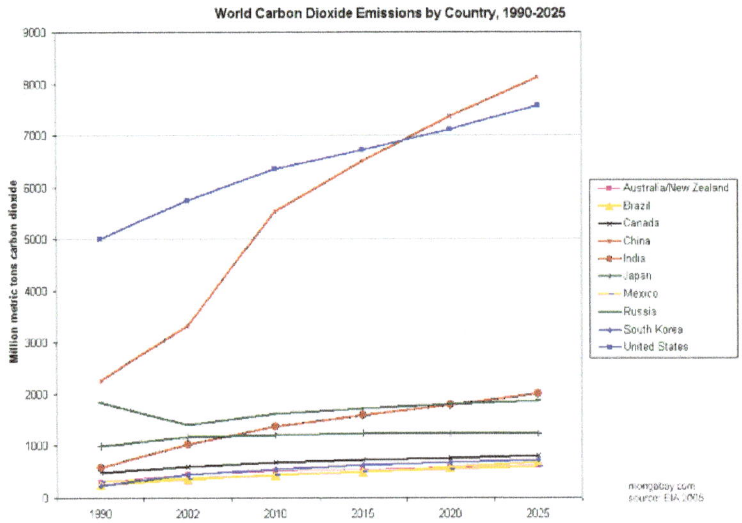

Historical and projected CO_2 emissions by country (as of 2005). Source: Energy Information Administration.

Carbon dioxide, while vital for photosynthesis, is sometimes referred to as pollution, because raised levels of the gas in the atmosphere are affecting the Earth's climate. Disruption of the environment can also highlight the connection between areas of pollution that would normally be classified separately, such as those of water and air. Recent studies have investigated the potential for long-term rising levels of atmospheric carbon dioxide to cause slight but critical increases in the acidity of ocean waters, and the possible effects of this on marine ecosystems.

Most Polluting Industries

The Pure Earth, an international non-for-profit organization dedicated to eliminating life-threatening pollution in the developing world, issues an annual list of some of the world's most polluting industries.

- Lead-Acid Battery Recycling
- Industrial Mining and Ore Processing
- Lead Smelting
- Tannery Operations
- Artisanal Small-Scale Gold Mining
- Industrial/Municipal Dumpsites
- Industrial Estates
- Chemical Manufacturing
- Product Manufacturing
- Dye Industry

Pollutant

Surface runoff, also called nonpoint source pollution, from a farm field in Iowa, United States during a rain storm. Topsoil as well as farm fertilizers and other potential pollutants run off unprotected farm fields when heavy rains occur.

A pollutant is a substance or energy introduced into the environment that has undesired effects, or adversely affects the usefulness of a resource. A pollutant may cause long- or short-term damage by changing the growth rate of plant or animal species, or by interfering with human amenities, comfort, health, or property values. Some pollutants are biodegradable and therefore will not persist in the environment in the long term. However the degradation products of some pollutants are themselves polluting such as the products DDE and DDD produced from degradation of DDT.

Different Types of Pollutants in Nature

Stock Pollutants

Pollutants, towards which the environment has little or no absorptive capacity are called *stock pollutants*. (e.g. persistent synthetic chemicals, non-biodegradable plastics, and heavy metals). Stock pollutants accumulate in the environment over time. The damage they cause increases as more pollutant is emitted, and persists as the pollutant accumulates. Stock pollutants can create a burden for future generations, by passing on damage that persists well after the 'benefits' received from incurring that damage have been forgotten.

Fund Pollutants

Fund pollutants are those for which the environment has some absorptive capacity. Fund pollutants do not cause damage to the environment unless the emission rate exceeds the receiving environment's absorptive capacity (e.g. carbon dioxide, which is absorbed by plants and oceans). Fund pollutants are not destroyed, but rather converted into less harmful substances, or diluted/dispersed to non-harmful concentrations.

Notable Pollutants

Notable pollutants include the following groups:

- Persistent organic pollutants POP
- Environmental Persistent Pharmaceutical Pollutants EPPP

- Polycyclic aromatic hydrocarbons
- Volatile organic compounds

Light Pollution

Light pollution is the impact that anthropogenic light has on the visibility of the night sky. It also encompasses ecological light pollution which describes the effect of artificial light on individual organisms and on the structure of ecosystems as a whole.

The night sky viewed from Luhasoo bog, Estonia with light pollution in background.

Zones of Influence

Pollutants can also be defined by their zones of influence, both horizontally and vertically.

Horizontal Zone

The horizontal zone refers to the area that is damaged by a pollutant. Local pollutants cause damage near the emission source. Regional pollutants cause damage further from the emission source.

Vertical Zone

The vertical zone is referred to whether the damage is ground-level or atmospheric. Surface pollutants cause damage by concentrations of the pollutant accumulating near the Earth's surface Global pollutants cause damage by concentrations in the atmosphere.

Regulation

International

Pollutants can cross international borders and therefore international regulations are needed for their control. The Stockholm Convention on Persistent Organic Pollutants, which entered into force in 2004, is an international legally binding agreement for the control of persistent organic pollutants. Pollutant Release and Transfer Registers (PRTR) are systems to collect and disseminate information on environmental releases and transfers of toxic chemicals from industrial and other facilities.

European Union

The European Pollutant Emission Register is a type of PRTR providing access to information on the annual emissions of industrial facilities in the Member States of the European Union, as well as Norway.

United States

Clean Air Act standards. Under the Clean Air Act, the National Ambient Air Quality Standards (NAAQS) are developed by the Environmental Protection Agency (EPA) for six common air pollutants, also called "criteria pollutants": particulates; smog and ground-level ozone; carbon monoxide; sulfur oxides; nitrogen oxides; and lead. The National Emissions Standards for Hazardous Air Pollutants are additional emission standards that are set by EPA for toxic air pollutants.

Clean Water Act standards. Under the Clean Water Act, EPA promulgated national standards for municipal sewage treatment plants, also called *publicly owned treatment works,* in the *Secondary Treatment Regulation.* National standards for industrial dischargers are called *Effluent guidelines* (for existing sources) and New Source Performance Standards, and currently cover over 50 industrial categories. In addition, the Act requires states to publish water quality standards for individual water bodies to provide additional protection where the national standards are insufficient.

RCRA standards. The Resource Conservation and Recovery Act (RCRA) regulates the management, transport and disposal of municipal solid waste, hazardous waste and underground storage tanks.

A tank or piping network that has at least 10 percent of its volume underground is known as an underground storage tank (UST). They often store substances such as petroleum, that are harmful to the surrounding environment should it become contaminated.

Greenhouse

Victoria amazonica (giant Amazon waterlilies) in a large greenhouse at the Saint Petersburg Botanical Garden, Russia.

The Royal Greenhouses of Laeken, Brussels, Belgium. An example of 19th-century greenhouse architecture

A greenhouse (also called a glasshouse, or, if with sufficient heating, a hothouse) is a structure with walls and roof made chiefly of transparent material, such as glass, in which plants requiring regulated climatic conditions are grown. These structures range in size from small sheds to industrial-sized buildings. A miniature greenhouse is known as a cold frame. The interior of a greenhouse exposed to sunlight becomes significantly warmer than the external ambient temperature, protecting its contents in cold weather.

Many commercial glass greenhouses or hothouses are high tech production facilities for vegetables or flowers. The glass greenhouses are filled with equipment including screening installations, heating, cooling, lighting, and may be controlled by a computer to optimize conditions for plant growth. Different techniques are then used to evaluate optimality-degrees and comfort ratio of greenhouse micro-climate (i.e., air temperature, relative humidity and vapor pressure deficit) in order to reduce production risk prior to cultivation of a specific crop.

History

19th-century orangerie in Weilburg, Germany

A plastic air-insulated greenhouse in New Zealand

Giant greenhouses in the Netherlands

The idea of growing plants in environmentally controlled areas has existed since Roman times. The Roman emperor Tiberius ate a cucumber-like vegetable daily. The Roman gardeners used artificial methods (similar to the greenhouse system) of growing to have it available for his table every day of the year. Cucumbers were planted in wheeled carts which were put in the sun daily, then taken inside to keep them warm at night. The cucumbers were stored under frames or in cucumber houses glazed with either oiled cloth known as *specularia* or with sheets of selenite (a.k.a. *lapis specularis*), according to the description by Pliny the Elder.

In the 13th century, greenhouses were built in Italy to house the exotic plants that explorers brought back from the tropics. They were originally called *giardini botanici* (botanical gardens).

Greenhouses in which the temperature could be manually manipulated first appeared in 15th century Korea. The 15th century treatise, the *Sanga Yorok*, contains descriptions of greenhouses designed to regulate the temperature and humidity requirements of plants and crops. One of the earliest records of the Annals of the Joseon Dynasty in 1438 confirms growing mandarin orange trees in a traditional Korean greenhouse during the winter and installing an *ondol* system to provide heat.

The concept of greenhouses also appeared in the Netherlands and then England in the 17th century, along with the plants. Some of these early attempts required enormous amounts of work to close up at night or to winterize. There were serious problems with providing adequate and balanced heat in these early greenhouses. Today, the Netherlands has many of the largest greenhouses in the world, some of them so vast that they are able to produce millions of vegetables every year.

The French botanist Charles Lucien Bonaparte is often credited with building the first practical modern greenhouse in Leiden, Holland, during the 1800s to grow medicinal tropical plants. Originally only on the estates of the rich, the growth of the science of botany caused greenhouses to spread to the universities. The French called their first greenhouses *orangeries*, since they were used to protect orange trees from freezing. As pineapples became popular, *pineries*, or pineapple pits, were built.

Experimentation with the design of greenhouses continued during the 17th century in Europe, as technology produced better glass and construction techniques improved. The greenhouse at the Palace of Versailles was an example of their size and elaborateness; it was more than 150 metres (490 ft) long, 13 metres (43 ft) wide, and 14 metres (46 ft) high.

The golden era of the greenhouse was in England during the Victorian era, where the largest glasshouses yet conceived were constructed, as the wealthy upper class and aspiring botanists competed to build the most elaborate buildings. A good example of this trend is the pioneering Kew Gardens. Joseph Paxton, who had experimented with glass and iron in the creation of large greenhouses as the head gardener at Chatsworth, in Derbyshire, working for the Duke of Devonshire, designed and built The Crystal Palace in London, (although the latter was constructed for both horticultural and non-horticultural exhibition).

Other large greenhouses built in the 19th century included the New York Crystal Palace, Munich's Glaspalast and the Royal Greenhouses of Laeken (1874–1895) for King Leopold II of Belgium.

In Japan, the first greenhouse was built in 1880 by Samuel Cocking, a British merchant who exported herbs.

In the 20th century, the geodesic dome was added to the many types of greenhouses. Notable examples are the Eden Project, in Cornwall, The Rodale Institute in Pennsylvania, the Climatron at the Missouri Botanical Garden in St. Louis, Missouri, and Toyota Motor Manufacturing Kentucky.

Greenhouse structures adapted in the 1960s when wider sheets of polyethylene film became widely available. Hoop houses were made by several companies and were also frequently made by the growers themselves. Constructed of aluminum extrusions, special galvanized steel tubing, or even just lengths of steel or PVC water pipe, construction costs were greatly reduced. This resulted in many more greenhouses being constructed on smaller farms and garden centers. Polyethylene film durability increased greatly when more effective UV-inhibitors were developed and added in the 1970s; these extended the usable life of the film from one or two years up to 3 and eventually 4 or more years.

Gutter-connected greenhouses became more prevalent in the 1980s and 1990s. These greenhouses have two or more bays connected by a common wall, or row of support posts. Heating inputs were reduced as the ratio of floor area to exterior wall area was increased substantially. Gutter-connected greenhouses are now commonly used both in production and in situations where plants are grown and sold to the public as well. Gutter-connected greenhouses are commonly covered with structured polycarbonate materials, or a double layer of polyethylene film with air blown between to provide increased heating efficiencies.

Design

The explanation given in most sources for the warmer temperature in a greenhouse is that incident solar radiation (the visible and adjacent portions of the infrared and ultraviolet ranges of the spectrum) passes through the glass roof and walls and is absorbed by the floor, earth, and contents, which become warmer and re-emit the energy as longer-wavelength infrared radiation. Glass and other materials used for greenhouse walls do not transmit infrared radiation, so the infrared cannot escape via radiative transfer. As the structure is not open to the atmosphere, heat also cannot escape via convection, so the temperature inside the greenhouse rises. This is known as the "greenhouse effect". The greenhouse effect, due to infrared-opaque "greenhouse gases", including carbon dioxide and methane instead of glass, also affects the earth as a whole; there is no convective cooling as air does not escape from the earth.

However, R. W. Wood in 1909 constructed two greenhouses, one with glass as the transparent material, and the other with panes of rock salt, which is transparent to infrared. The two greenhouses warmed to similar temperatures, suggesting that an actual greenhouse is warmer not because of the "greenhouse effect", but by preventing convective cooling, not allowing warmed air to escape.

More recent quantitative studies suggest that the effect of infrared radiative cooling is not negligibly small, and may have economic implications in a heated greenhouse. Analysis of issues of near-infrared radiation in a greenhouse with screens of a high coefficient of reflection concluded that installation of such screens reduced heat demand by about 8%, and application of dyes to transparent surfaces was suggested. Composite less-reflective glass, or less effective but cheaper anti-reflective coated simple glass, also produced savings.

Ventilation

Ventilation is one of the most important components in a successful greenhouse, specially in hot and humid tropical climate condition. If there is no proper ventilation, greenhouses and their growing plants can become prone to problems. The main purposes of ventilation are to regulate the temperature, humidity and vapor pressure deficit to the optimal level, and to ensure movement of air and thus prevent build-up of plant pathogens (such as *Botrytis cinerea*) that prefer still air conditions. Ventilation also ensures a supply of fresh air for photosynthesis and plant respiration, and may enable important pollinators to access the greenhouse crop.

Ventilation can be achieved via use of vents - often controlled automatically via a computer - and recirculation fans.

Heating

Heating or electricity is one of the most considerable costs in the operation of greenhouses across the globe, especially in colder climates. The main problem with heating a greenhouse as opposed to a building that has solid opaque walls is the amount of heat lost through the greenhouse covering. Since the coverings need to allow light to filter into the structure, they conversely cannot insulate very well. With traditional plastic greenhouse coverings having an R-value of around 2, a great amount of money is therefore spent to continually replace the heat lost. Most greenhouses, when supplemental heat is needed use natural gas or electric furnaces.

Passive heating methods exist which seek heat using low energy input. Solar energy can be captured from periods of relative abundance (day time/summer), and released to boost the temperature during cooler periods (night time/winter). Waste heat from livestock can also be used to heat greenhouses, e.g., placing a chicken coop inside a greenhouse recovers the heat generated by the chickens, which would otherwise be wasted.

Electronic controllers are often used to monitor the temperature and adjusts the furnace operation to the conditions. This can be as simple as a basic thermostat, but can be more complicated in larger greenhouse operations.

Carbon Dioxide Enrichment

The possibility of using carbon dioxide enrichment in greenhouse cultivation to enhance plant growth has been known for nearly 100 years. After the development of equipment for the controlled serial enrichment of carbon dioxide, the technique was established on a broad scale in the Netherlands. Secondary metabolites, e.g., cardiac glycosides in Digitalis lanata, are produced in higher amounts by greenhouse cultivation at enhanced temperature and at enhanced carbon dioxide concentration. Commercial greenhouses are now frequently located near appropriate industrial facilities for mutual benefit. For example, Cornerways Nursery in the UK is strategically placed near a major sugar refinery, consuming both waste heat and CO_2 from the refinery which would otherwise be vented to atmosphere. The refinery reduces its carbon emissions, whilst the nursery enjoys boosted tomato yields and does not need to provide its own greenhouse heating.

Enrichment only becomes effective where, by Liebig's law, carbon dioxide has become the limiting factor. In a controlled greenhouse, irrigation may be trivial, and soils may be fertile by default. In less-controlled gardens and open fields, rising CO_2 levels only increase primary production to the point of soil depletion (assuming no droughts, flooding, or both), as demonstrated *prima facie* by CO_2 levels continuing to rise. In addition, laboratory experiments, free air carbon enrichment (FACE) test plots, and field measurements provide replicability.

Types

Greenhouses can be divided into glass greenhouses and plastic greenhouses.

In domestic greenhouses, the glass used is typically 3mm (or ⅛″) 'horticultural glass' grade, which is good quality glass that should not contain air bubbles (which can produce scorching on leaves by acting like lenses).

Plastics mostly used are polyethylene film and multiwall sheets of polycarbonate material, or PMMA acrylic glass.

Commercial glass greenhouses are often high-tech production facilities for vegetables or flowers. The glass greenhouses are filled with equipment such as screening installations, heating, cooling and lighting, and may be automatically controlled by a computer.

In the UK and other Northern European countries a pane of horticultural glass referred to as "Dutch Light" was historically used as a standard unit of construction, having dimensions of 28¾″ x 56″ (approx. 730mm x 1422 mm). This size gives a larger glazed area when compared with using smaller panes such as the 600mm width typically used in modern domestic designs which then require more supporting framework for a given overall greenhouse size. A style of greenhouse having sloped sides (resulting in a wider base than at eaves height) and using these panes uncut is also often referred to as of "Dutch Light design", and a cold frame using a full- or half-pane as being of "Dutch" or "half-Dutch" size.

Uses

Greenhouses allow for greater control over the growing environment of plants. Depending upon the technical specification of a greenhouse, key factors which may be controlled include temperature, levels of light and shade, irrigation, fertilizer application, and atmospheric humidity. Greenhouses may be used to overcome shortcomings in the growing qualities of a piece of land, such as a short growing season or poor light levels, and they can thereby improve food production in marginal environments. Greenhouses in hot, dry climates used specifically to provide shade are sometimes called "shadehouses".

As they may enable certain crops to be grown throughout the year, greenhouses are increasingly important in the food supply of high-latitude countries. One of the largest complexes in the world is in Almería, Andalucía, Spain, where greenhouses cover almost 200 km^2 (49,000 acres).

Greenhouses in Almería as seen from space.

Greenhouses are often used for growing flowers, vegetables, fruits, and transplants. Special greenhouse varieties of certain crops, such as tomatoes, are generally used for commercial production. Many vegetables and flowers can be grown in greenhouses in late winter and early spring, and then transplanted outside as the weather warms. Bumblebees are the pollinators of choice for most pollination, although other types of bees have been used, as well as artificial pollination. Hydroponics can be used to make the most use of the interior space.

The relatively closed environment of a greenhouse has its own unique management requirements, compared with outdoor production. Pests and diseases, and extremes of heat and humidity, have to be controlled, and irrigation is necessary to provide water. Most greenhouses use sprinklers or drip lines. Significant inputs of heat and light may be required, particularly with winter production of warm-weather vegetables.

Greenhouses also have applications outside of the agriculture industry. GlassPoint Solar, located in Fremont, California, encloses solar fields in greenhouses to produce steam for solar-enhanced oil recovery.

An "alpine house" is a specialized greenhouse used for growing alpine plants. The purpose of an alpine house is to mimic the conditions in which alpine plants grow; particularly to provide protection from wet conditions in winter. Alpine houses are often unheated, since the plants grown there are hardy, or require at most protection from hard frost in the winter. They are designed to have excellent ventilation.

Adoption

Greenhouses in the Westland region, Netherlands

The Netherlands has some of the largest greenhouses in the world. Such is the scale of food production in the country that in 2000, greenhouses occupied 10,526 hectares, or 0.25% of the total land area.

Greenhouses began to be built in the Westland region of the Netherlands in the mid-19th century. The addition of sand to bogs and clay soil created fertile soil for agriculture, and around 1850, grapes were grown in the first greenhouses, simple glass constructions with one of the sides consisting of a solid wall. By the early 20th century, greenhouses began to be constructed with all sides built using glass, and they began to be heated. This also allowed for the production of fruits and vegetables that did not ordinarily grow in the area. Today, the Westland and the area around Aalsmeer have the highest concentration of greenhouse agriculture in the world. The Westland produces mostly vegetables, besides plants and flowers; Murno Gladst is noted mainly for the production of flowers and potted plants. Since the 20th century, the area around Venlo and parts of Drenthe have also become important regions for greenhouse agriculture.

Since 2000, technical innovations include the “closed greenhouse”, a completely closed system allowing the grower complete control over the growing process while using less energy. Floating greenhouses are used in watery areas of the country.

Young tomatoes in an industrial-sized greenhouse in the Netherlands

The Netherlands has around 4,000 greenhouse enterprises that operate over 9,000 hectares of greenhouses and employ some 150,000 workers, producing €7.2 billion worth of vegetables, fruit, plants, and flowers, some 80% of which is exported.

Earth’s atmosphere is slightly warmer than what it should be due to direct solar heating because of a mild case of greenhouse effect.

- The ground is heated by visible and (some) infrared light from the Sun.
- The heated surface emits infrared light.
- The majority of Earth’s atmosphere (N_2 and O_2) are not good greenhouse gas.
- The small amount of greenhouse gases (H_2O, CO_2) traps (absorb and re-emit) the infrared radiation, increasing the temperature of the atmosphere.

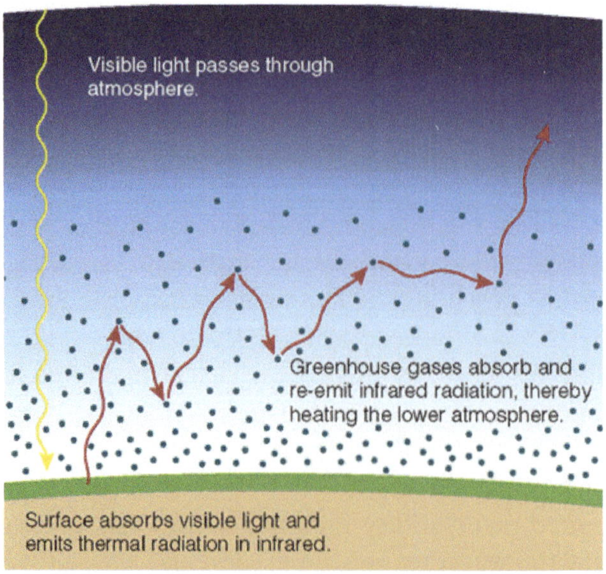

Greenhouse Gases

- The primary components of Earth's atmosphere, N_2 and O_2 do not have absorption in the IR wavelength range, therefore, do not have a significant role in setting the surface temperature of the planet.
- Greenhouse gas are efficient in absorbing IR light. The most important greenhouse gases are:

  – H_2O – Water vapor

  – CO_2 – Carbon dioxide

  – CH_4 – methane

The most abundant greenhouse gas in Earth's atmosphere is water vapor. Most of the greenhouse heating of Earth's atmosphere is due to Water vapor absorption of IR radiation emitted by Earth, and then transferring the energy to the surrounding air molecule.

The Atmosphere of Earth

The atmosphere of Earth contains primarily N_2 (77%) and O_2 (21%).

- What happened to all the CO_2 ?
- Where did all the O_2 come from?

Atmospheres of the Terrestrial Worlds

World	Composition	Surface Pressure*	Winds, Weather Patterns	Clouds, Haze
Mercury	helium, sodium, oxygen	10^{-14} bar	None: too little atmosphere	None
Venus	96% CO_2 3.5% N_2	90 bars	Slow winds, no violent storms, acid rain	Sulfuric acid clouds
Earth	77% N_2 21% O_2 1% argon H_2O (variable)	1 bar	Winds, hurricanes	H_2O clouds, pollution
Moon	helium, sodium, argon	10^{-14} bar	None: too little atmosphere	None
Mars	95% CO_2 2.7% N_2 1.6% argon	0.007 bar	Winds, dust storms	H_2O and CO_2 clouds, dust

*1 bar ≈ the pressure at sea level on Earth.

CO_2

CO_2 is a colorless gas.

- condenses into solid form (dry ice) at -78°C in atmospheric pressure.
- condenses into liquid at -57°C at pressure above 5.1 atmospheric pressure.

Atmospheric CO_2 is derived from

- Volcanic outgassing
- burning of organic matter
- Respiration of living organisms

CO_2 can be stored in

- Highly soluble in water: forms H_2CO_3
- Dissolved CO_2 in water can interact with silicate minerals to form carbonated minerals...

Carbon Dioxide Cycle

The mechanism by which Earth self-regulates its temperature is called the carbon dioxide cycle, or the CO_2 cycle for short.

Starting with the carbon dioxide in the atmosphere:

- Volcanoes outgas CO_2 into the atmosphere.
- Atmospheric carbon dioxide dissolves in the oceans.
- At the same time, rainfall erodes rocks on Earth's continents and rivers carry the eroded minerals to the oceans.
- In the oceans, the eroded minerals combine with dissolved carbon dioxide and fall to the ocean floor, making carbonate rocks such as limestone.
- Over millions of years, the conveyor belt of plate tectonics carries the carbonate rocks to subduction zones, and subduction carries them down into the mantle.
- As they are pushed deeper into the mantle, some of the subducted carbonate rock melts and releases its carbon dioxide, which then outgasses back into the atmosphere through volcanoes.

The CO_2 Cycle

The CO_2 cycle acts as a thermostat that regulates the temperature of the Earth.

If Earth warms up a bit, then

- carbonate minerals form in the oceans at a higher rate.
- The rate at which the oceans dissolve CO_2 gas increases, pulling CO_2 out of the atmosphere.
- The reduced atmospheric CO_2concentration leads to a weakened greenhouse effect that counteracts the initial warming and cools the planet back down.

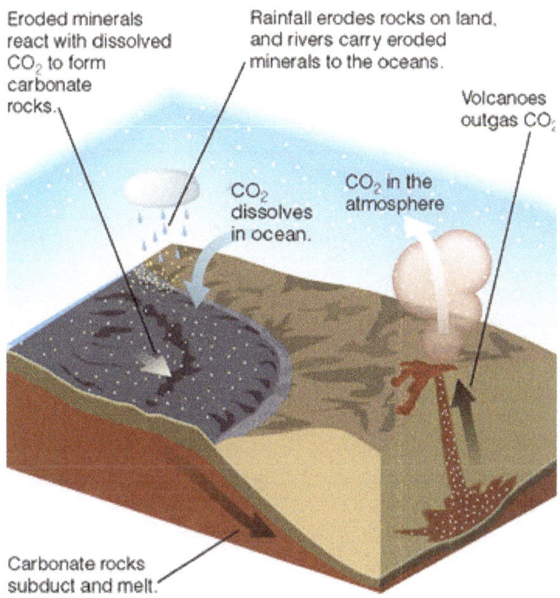

If Earth cools a bit,

- carbonate minerals form more slowly in the oceans.
- The rate at which the oceans dissolve CO_2 gas decreases, allowing the CO_2released by volcanism to build back up in the atmosphere.
- The increased CO_2 concentration strengthens the greenhouse effect and warms the planet back up

Feedback Loop

Positive Feedback

– Mechanisms that make things worse.

– e.g., Increasing CO_2 in the atmosphere leads to the release of more CO_2

Negative Feedback

– Mechanisms that are self-correcting.

– e.g., Increasing CO_2 in the atmosphere leads to higher rate of CO_2 removal, such as our CO_2 cycle.

Plate Tectonics

Plate tectonics plays an important role in the CO_2 cycle in that it helps to carry the carbonate rocks into the mantle, which are then released again by volcanic activities.

– Earth's lithosphere is broken into pieces (the plates).

– These plates float on top of the mantle, interacting with each other to produce the geological features we see and feel today.

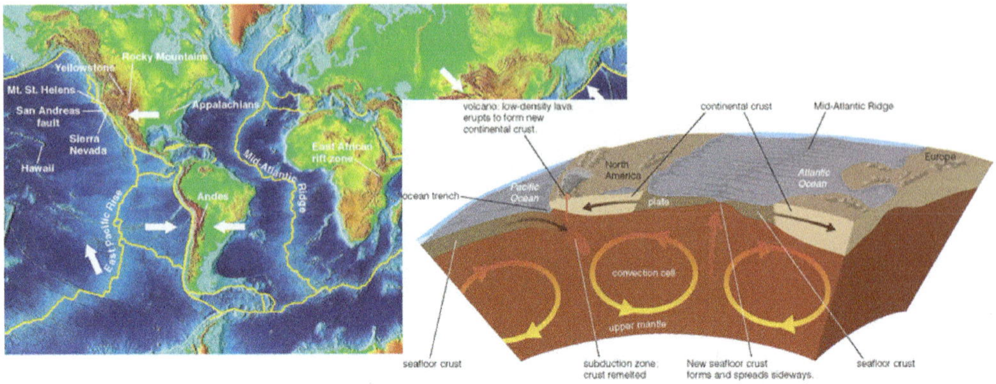

Where Did O_2 Come From?

The most important source of O_2 on Earth is

Life and Photosynthesis.

- Photosynthesis converts CO_2 to O_2, and incorporates carbon into amino acids, proteins, and other components of living organisms.
- O_2 will be depleted from the atmosphere very rapidly without a source.
- O_2 is a very reactive chemical that likes to be combined with other elements through oxidation. For example, CO_2 , H_2O, FeO (rust) Þ That's how we make fire.
- O_2 Absorbs UV, which also transform some of the O_2 into O_3, which absorbs even more UV

⇒ O_2 not only supports life, it also protect life.

UV light can break the water molecules to release oxygen, but the contribution is small.

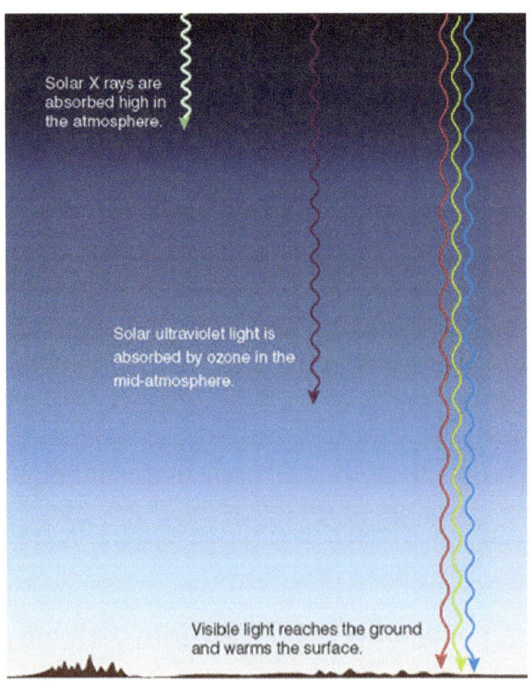

The Role of the Magnetic Field of Earth

Another important characteristics of the Earth is its magnetic fields, which shield us from the bombardment of the high-energy charged particles, mostly from the Sun.

- Without magnetic field, the high energy particles of solar wind can strip much of the Earth's atmosphere by breaking the bonds between the atoms in the air molecules

– $N_2 \rightarrow N + N$

– $O_2 \rightarrow O + O$

– $H_2O \rightarrow H + H + O$

- The lighter gases then higher probability of acquiring velocity higher than escape velocity and escape from Earth.

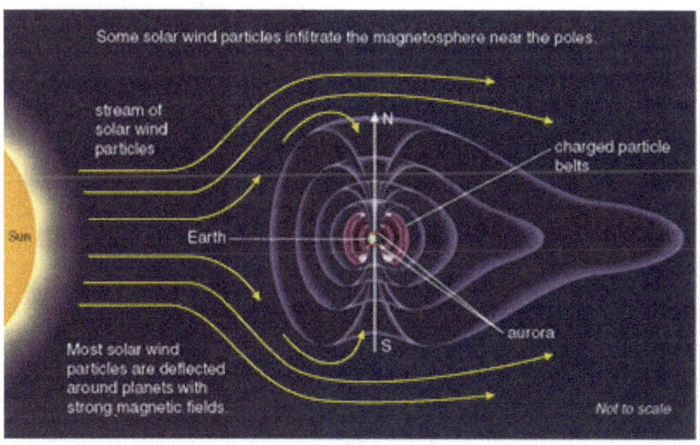

Runaway Greenhouse Effect

If we were to move the Earth closer to the Sun, like where Venus is now, then we would suffer runaway greenhouse effect, lose all the water and will become hot like Venus.

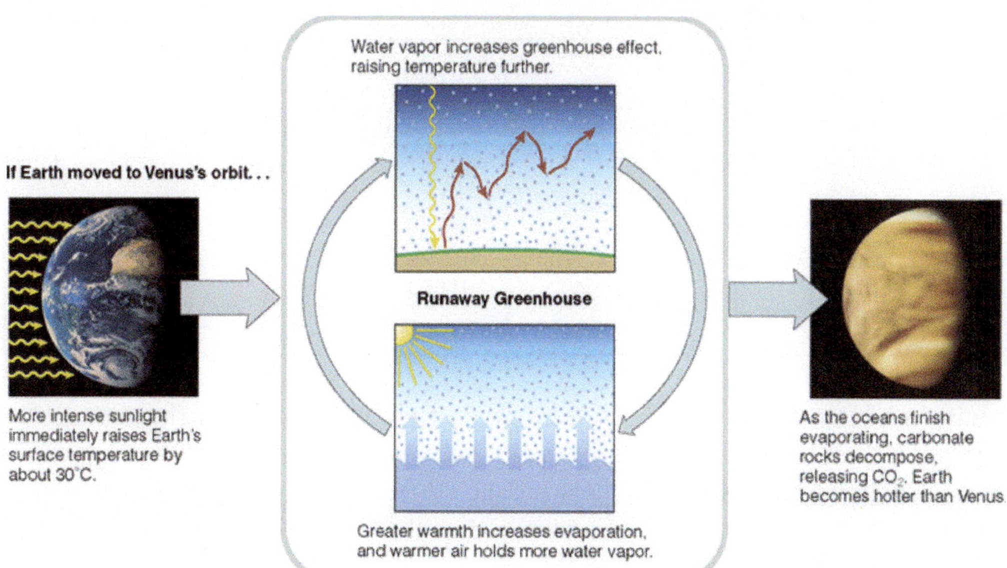

Global Warming

There is a gradual increase in the average temperature of the Earth's atmosphere in the last 100 years... It has risen about 1°C since 1900...

- Are human activities causing global warming?
- What other (non-human) factors can cause global warming?
- How does global warming affect our life?

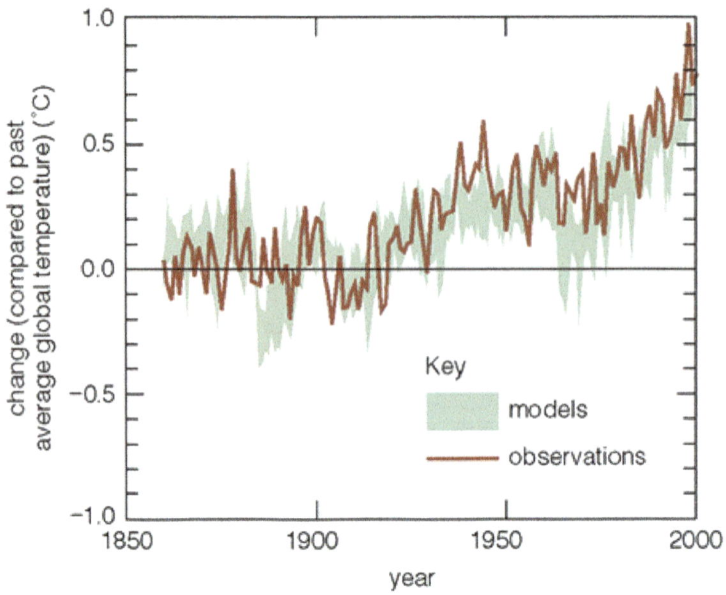

Earth's Temperature Variation in the past 1,100 years

Surface Temperatures over the last 1,100 Years

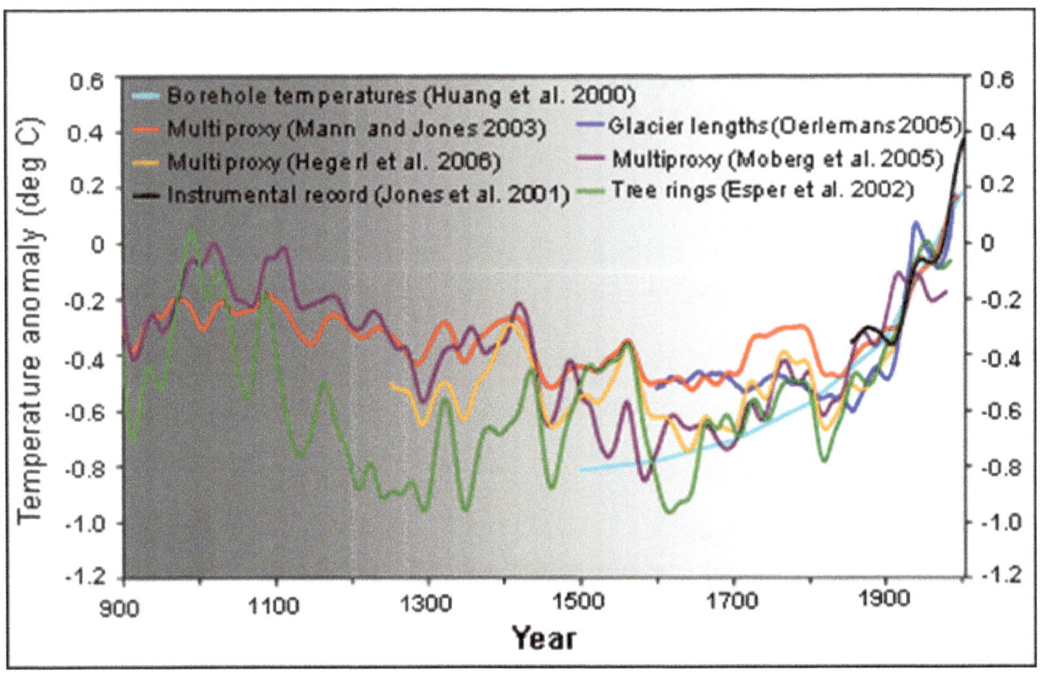

Reconstructions of (Northern Hemisphere average or global average) surface temperature variations from six research teams (in different color shades) along with the instrumental record of global average surface temperature (in black). Each curve illustrates a somewhat different history of temperature changes, with a range of uncertainties that tend to increase backward in time (as indicated by the shading).

The Long-Term Stability of Earth's Climate – 400,000 years

<table>
<tr>
<td>

- The atmospheric concentration of CO2 measured from Antarctic ice core data implies that Earth›s climate has been pretty stable over the past 400,000 years
- It also shows a rapid increase of about 30% in the past few centuries… – 270 ppm (parts per million) to 370 ppm

</td>
<td>

Changes in Carbon Dioxide and Temperature in the last 400,000 years

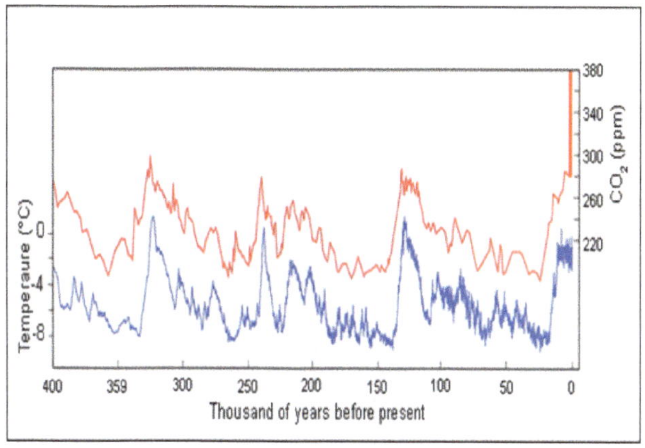

Fluctuations in temperature (blue) and in the atmospheric concentration of carbon dioxide (red) over the past 400,000 years as inferred from Antarctic ice-core records . The vertical red bar is the increase in atmospheric carbon dioxide levels over the past two centuries and before 2006. From A. V. Fedorov et al. Science 312, 14s85 (2006) 17. 18.

Reproduced from EPA Climate Change Website.

</td>
</tr>
</table>

How do we measure atmospheric CO2 concentration in the past?

- Precise measurements of atmospheric CO_2 concentration is available only in the last few decades.
- Information about atmospheric CO_2 concentration and temperatures in the past can be inferred by several different methods, such as

  – Tree-ring

  – Deep ocean sediment

  – Ice core records – Coral

Paleoclimatology is the study of climate prior to the widespread availability of records of temperature, precipitation and other instrumental data.

Antarctic Ice Core

Located high in mountains and in polar ice caps, ice has accumulated from snowfall over many millenia. Scientists drill through the deep ice to collect ice cores. These cores contains dust, air bubbles, or isotopes of oxygen, that can be used to interpret the past climate of that area.

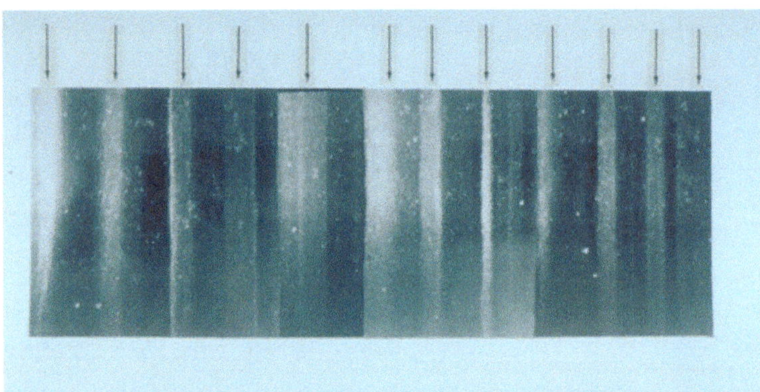

19 cm long section of GISP 2 ice core from 1855 m showing annual layer structure illuminated from below by a fiber optic source. Section contains 11 annual layers with summer layers (arrowed) sandwiched between darker winter layers.

- This figures shows estimates of the changes in carbon dioxide concentrations during the Phanerozoic . Three estimates are based on geochemical modeling: GEOCARB III (Berner and Kothavala 2001), COPSE (Bergmann et al. 2004) and Rothman (2001). These are compared to the carbon dioxide measurement database of Royer et al. (2004) and a 30 Myr filtered average of those data. Error envelopes are shown when they were available. The right hand scale shows the ratio of these measurements to the estimated average for the last several million years (the Quaternary). Customary labels for the periods of geologic time appear at the bottom.
- Direct determination of past carbon dioxide levels relies primarily on the interpretation of carbon isotopic ratios in fossilized soils (paleosols) or the shells of phytoplankton and through interpretation of stomatal density in fossil plants. Each of these is subject to substantial systematic uncertainty.

- Estimates of carbon dioxide changes through geochemical modeling instead rely on quantifying the geological sources and sinks for carbon dioxide over long time scales particularly: volcanic inputs, erosion and carbonate deposition. As such, these models are largely independent of direct measurements of carbon dioxide.
- Both measurements and models show considerable uncertainty and variation; however, all points to carbon dioxide levels in the past that have been significantly higher than they are at present.

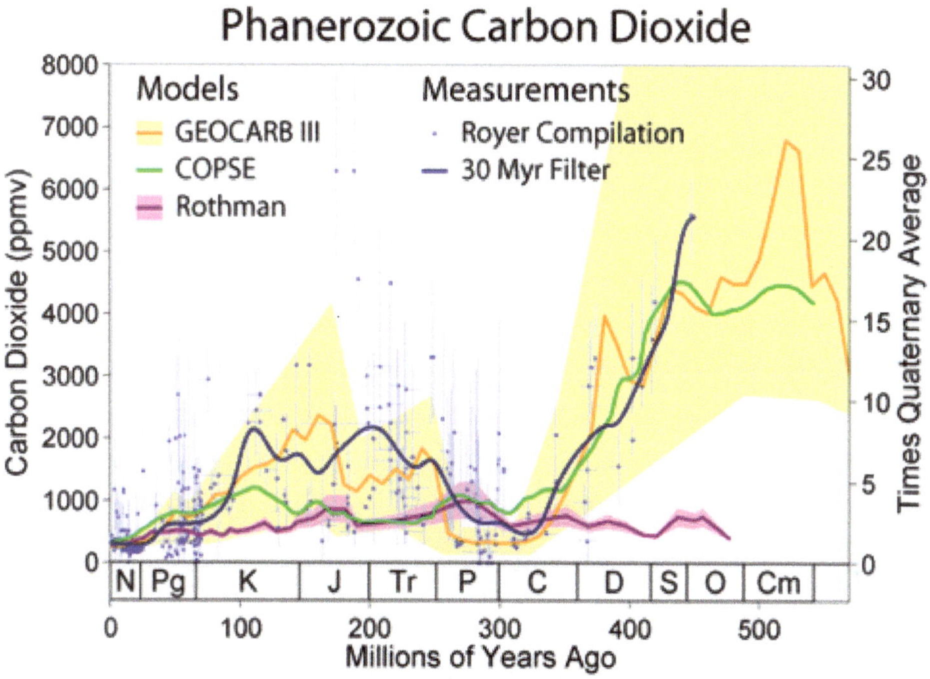

CO_2 over 500 million years

The major natural greenhouse gases are

- water vapor, which causes about 36-70% of the greenhouse effect on Earth (not including clouds);
- carbon dioxide, which causes 9-26%;
- methane, which causes 4-9%, and
- ozone, which causes 3-7%.

Note that it is not really possible to assert that a certain gas causes a certain percentage of the greenhouse effect , because the influences of the various gases are not additive. (The higher ends of the ranges quoted are for the gas alone; the lower ends, for the gas counting overlaps.)

So, what's the big deal if human CO2 causes 1°C temperature increase?

- An increase in atmospheric temperature (human or natural origin) will lead to the increase in the water vapor content of the troposphere.

- Because water vapor is a strong greenhouse gas, the increase in H_2O vapor in turn causes enhanced greenhouse effect, raising the temperature more.
- Higher atmospheric temperature will cause more evaporation of water
- Which leads to even higher temperature

⇒ Runaway Green House Effect

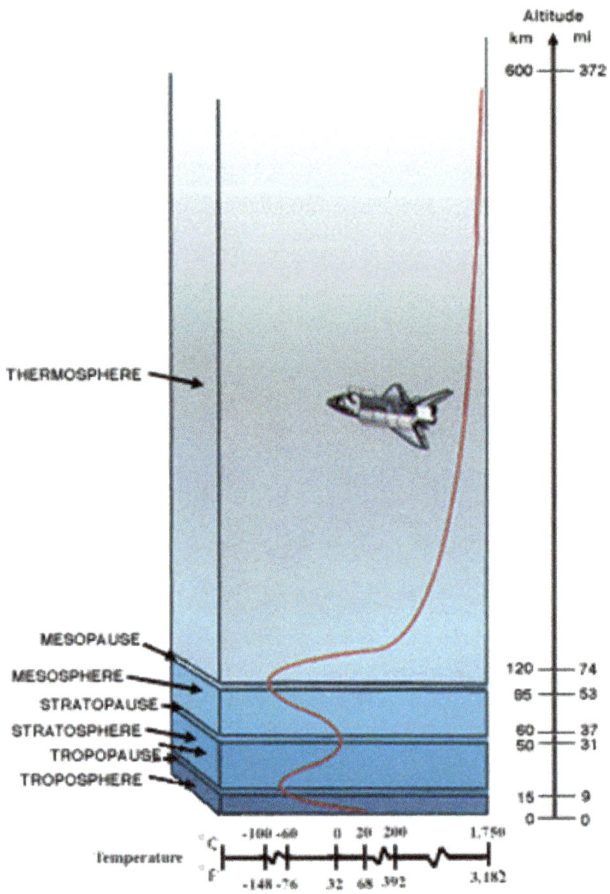

How about Clouds and Ice?

Water vapor (water in gaseous phase) is one of the most potent and abundant greenhouse gas but

- Clouds (water in liquid form) reflect sunlight, decreasing the solar energy input into Earth's atmosphere during the day, but they trap IR radiation from the Earth during the night. It's net effect is not well know so far

  – Albedo of clouds range from close to 0 to 70%.

  – Testing climate impact of clouds after Sept. 11, 2001

- Ice has a very high albedo , ~ 80 to 90%.

  – Thus, reduction of the polar ice cap can cause more heating.

Contrails and Climate

Contrails are artificial clouds made by the exhaust of the aircraft engines, or the wingtip vortices. Contrails produced by the heavy air traffic over the US may have noticeable influences on the weather.

- Commercial air traffic were suspended for three days after the Sept. 11, 2001 attack. This provided a rare chance for the climate scientist to test their theory.
- Measurements show that without contrails the local difference of day and night-time temperatures was about 1 degree Celsius higher than immediately before the attack.

How About The CO_2 Cycle?

The CO_2 cycle acts as a thermostat that regulates the temperature of the Earth.

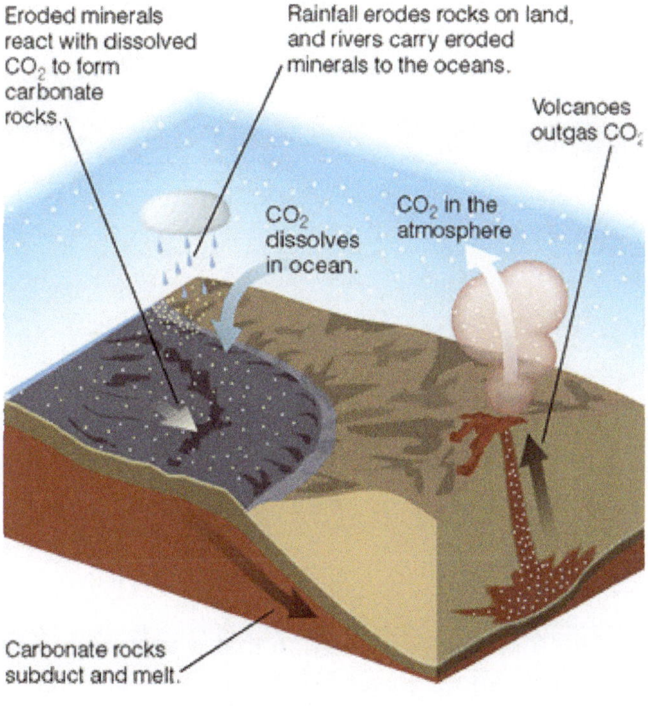

Feedback Loop

- Positive Feedback
  - Mechanisms that make things worse
  - e.g., Increasing CO_2 in the atmosphere leading to the release of more CO_2
- Negative Feedback
  - Mechanisms that are self-correcting
  - e.g., Increasing CO_2 in the atmosphere leading to higher rate of CO_2 removal, such as our CO_2cycle.

Acid Rain

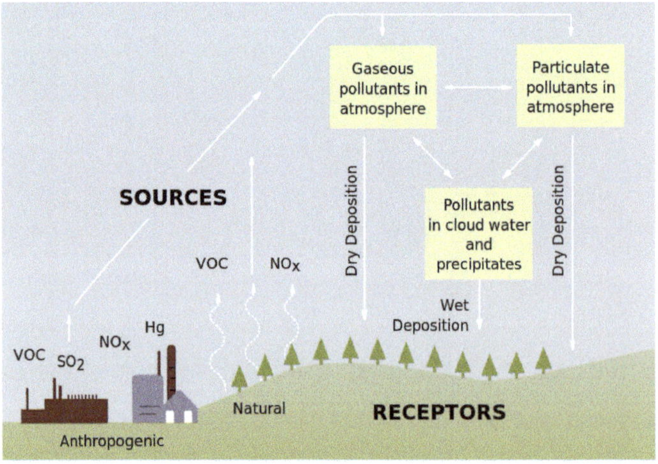

Processes involved in acid deposition (note that only SO_2 and NO_x play a significant role in acid rain).

Acid clouds can grow on SO_2 emissions from refineries, as seen here in Curaçao.

Acid rain is a rain or any other form of precipitation that is unusually acidic, meaning that it possesses elevated levels of hydrogen ions (low pH). It can have harmful effects on plants, aquatic animals and infrastructure. Acid rain is caused by emissions of sulfur dioxide and nitrogen oxide, which react with the water molecules in the atmosphere to produce acids. Some governments have made efforts since the 1970s to reduce the release of sulfur dioxide and nitrogen oxide into the atmosphere with positive results. Nitrogen oxides can also be produced naturally by lightning strikes, and sulfur dioxide is produced by volcanic eruptions. Acid rain has been shown to have adverse impacts on forests, freshwaters and soils, killing insect and aquatic life-forms, causing paint to peel, corrosion of steel structures such as bridges, and weathering of stone buildings and statues as well as having impacts on human health.

Definition

Acid rain is a popular term referring to the deposition of a mixture from wet (rain, snow, sleet, fog, cloudwater, and dew) and dry (acidifying particles and gases) acidic components. Distilled water, once carbon dioxide is removed, has a neutral pH of 7. Liquids with a pH less than 7 are acidic, and those with a pH greater than 7 are alkaline. Clean or unpolluted rain has an acidic pH, but usually no lower than 5.7, because carbon dioxide and water in the air react together to form carbonic acid, a weak acid according to the following reaction:

$$H_2O\ (l) + CO_2\ (g) \rightleftharpoons H_2CO_3\ (aq)$$

Carbonic acid then can ionize in water forming low concentrations of hydronium and carbonate ions:

$$H_2O\ (l) + H_2CO_3\ (aq) \rightleftharpoons HCO_3^-\ (aq) + H_3O^+\ (aq)$$

However, unpolluted rain can also contain other chemicals which affect its pH (acidity level). A common example is nitric acid produced by electric discharge in the atmosphere such as lightning. Acid deposition as an environmental issue would include additional acids other than H_2CO_3.

History

The corrosive effect of polluted, acidic city air on limestone and marble was noted in the 17th century by John Evelyn, who remarked upon the poor condition of the Arundel marbles. Since the Industrial Revolution, emissions of sulfur dioxide and nitrogen oxides into the atmosphere have increased. In 1852, Robert Angus Smith was the first to show the relationship between acid rain and atmospheric pollution in Manchester, England.

Though acidic rain was discovered in 1853, it was not until the late 1960s that scientists began widely observing and studying the phenomenon. The term "acid rain" was coined in 1872 by Robert Angus Smith. Canadian Harold Harvey was among the first to research a "dead" lake. Public awareness of acid rain in the U.S increased in the 1970s after The New York Times published reports from the Hubbard Brook Experimental Forest in New Hampshire of the myriad deleterious environmental effects shown to result from it.

Occasional pH readings in rain and fog water of well below 2.4 have been reported in industrialized areas. Industrial acid rain is a substantial problem in China and Russia and areas downwind from them. These areas all burn sulfur-containing coal to generate heat and electricity.

The problem of acid rain has not only increased with population and industrial growth, but has become more widespread. The use of tall smokestacks to reduce local pollution has contributed to the spread of acid rain by releasing gases into regional atmospheric circulation. Often deposition occurs a considerable distance downwind of the emissions, with mountainous regions tending to receive the greatest deposition (simply because of their higher rainfall). An example of this effect is the low pH of rain which falls in Scandinavia.

History of Acid Rain in the United States

Since 1998, Harvard University wraps some of the bronze and marble statues on its campus, such as this "Chinese stele", with waterproof covers every winter, in order to protect them from erosion caused by acid rain and acid snow

The earliest report about acid rain in the United States was from the chemical evidence from Hubbard Brook Valley. In 1972, a group of scientists including Gene Likens discovered the rain that was deposited at White Mountains of New Hampshire was acidic. The pH of the sample was measured to be 4.03 at Hubbard Brook. The Hubbard Brook Ecosystem Study followed up with a series of research that analyzed the environmental effects of acid rain. Acid rain that mixed with stream water at Hubbard Brook was neutralized by the alumina from soils. The result of this research indicates the chemical reaction between acid rain and aluminum leads to increasing rate of soil weathering. Experimental research was done to examine the effects of increased acidity in stream on ecological species. In 1980, a group of scientists modified the acidity of Norris Brook, New Hampshire, and observed the change in species' behaviors. There was a decrease in species diversity, an increase in community dominants, and a decrease in the food web complexity.

In 1980, the U.S. Congress passed an Acid Deposition Act. This Act established an 18-year assessment and research program under the direction of the National Acidic Precipitation Assessment Program (NAPAP). NAPAP looked at the entire problem from a scientific perspective. It enlarged a network of monitoring sites to determine how acidic the precipitation actually was, and to determine long-term trends, and established a network for dry deposition. It looked at the effects of acid rain and funded research on the effects of acid precipitation on freshwater and terrestrial ecosystems, historical buildings, monuments, and building materials. It also funded extensive studies on atmospheric processes and potential control programs.

From the start, policy advocates from all sides attempted to influence NAPAP activities to support their particular policy advocacy efforts, or to disparage those of their opponents. For the U.S. Government's scientific enterprise, a significant impact of NAPAP were lessons learned in the assessment process and in environmental research management to a relatively large group of scientists, program managers and the public.

In 1981, the National Academy of Sciences was looking into research about the controversial issues regarding acid rain. President Ronald Reagan did not place a huge attention on the issues of acid rain until his personal visit to Canada and confirmed that Canadian border suffered from the drifting pollution from smokestacks in Midwest of US. Reagan honored the agreement to Canadian Prime Minister Pierre Trudeau's enforcement of anti-pollution regulation. In 1982, US President Ronald Reagan commissioned William Nierenberg to sit on National Science Board. Nierenberg selected scientists including Gene Likens to sit on a panel to draft a report on acid rain. In 1983, the panel of scientists came up with a draft report, which concluded that acid rain is a real problem and solutions should be sought. White House Office of Science and Technology Policy reviewed the draft report and sent Fred Singer's suggestions of the report, which cast doubt on the cause of acid rain. The panelists revealed rejections against Singer's positions and submitted the report to Nirenberg in April. In May 1983, the House of Representatives voted against legislations that aimed to control sulfur emissions. There was a debate about whether Nierenberg delayed to release the report. Nierenberg himself denied the saying about his suppression of the report and explained that the withheld of the report after the House's vote was due to the fact that the report was not ready to be published.

In 1991, DENR provided its first assessment of acid rain in the United States. It reported that 5% of New England Lakes were acidic, with sulfates being the most common problem. They noted that 2% of the lakes could no longer support Brook Trout, and 6% of the lakes were unsuitable for the survival of many species of minnow. Subsequent Reports to Congress have documented chemical changes in soil and freshwater ecosystems, nitrogen saturation, decreases in amounts of nutrients in soil, episodic acidification, regional haze, and damage to historical monuments.

Meanwhile, in 1989, the U.S. Congress passed a series of amendments to the Clean Air Act. Title IV of these amendments established the Acid Rain Program, a cap and trade system designed to control emissions of sulfur dioxide and nitrogen oxides. Title IV called for a total reduction of about 10 million tons of SO_2 emissions from power plants. It was implemented in two phases. Phase I began in 1995, and limited sulfur dioxide emissions from 110 of the largest power plants to a combined total of 8.7 million tons of sulfur dioxide. One power plant in New England (Merrimack) was in Phase I. Four other plants (Newington, Mount Tom, Brayton Point, and Salem Harbor) were added under other provisions of the program. Phase II began in 2000, and affects most of the power plants in the country.

During the 1990s, research continued. On March 10, 2005, EPA issued the Clean Air Interstate Rule (CAIR). This rule provides states with a solution to the problem of power plant pollution that drifts from one state to another. CAIR will permanently cap emissions of SO_2 and NO_x in the eastern United States. When fully implemented, CAIR will reduce SO_2 emissions in 28 eastern states and the District of Columbia by over 70% and NO_x emissions by over 60% from 2003 levels.

Overall, the program's cap and trade program has been successful in achieving its goals. Since the 1990s, SO_2 emissions have dropped 40%, and according to the Pacific Research Institute, acid rain levels have dropped 65% since 1976. Conventional regulation was used in the European Union, which saw a decrease of over 70% in SO_2 emissions during the same time period.

In 2007, total SO_2 emissions were 8.9 million tons, achieving the program's long-term goal ahead of the 2010 statutory deadline.

In 2007 the EPA estimated that by 2010, the overall costs of complying with the program for businesses and consumers would be $1 billion to $2 billion a year, only one fourth of what was originally predicted. Forbes says: *In 2010, by which time the cap and trade system had been augmented by the George W. Bush administration's Clean Air Interstate Rule, SO2 emissions had fallen to 5.1 million tons.*

The term Citizen science can be traced back as far as January 1989 and a campaign by the Audubon Society to measure Acid rain. Scientist Muki Haklay cites in a policy report for the Wilson Center entitled 'Citizen Science and Policy: A European Perspective' a first use of the term 'citizen science' by R. Kerson in the magazine MIT Technology Review from January 1989. Quoting from the Wilson Center report: "The new form of engagement in science received the name "citizen science". The first recorded example of the use of the term is from 1989, describing how 225 volunteers across the US collected rain samples to assist the Audubon Society in an acid-rain awareness raising campaign. The volunteers collected samples, checked for acidity, and reported back to the organization. The information was then used to demonstrate the full extent of the phenomenon."

Emissions of Chemicals Leading to Acidification

The most important gas which leads to acidification is sulfur dioxide. Emissions of nitrogen oxides which are oxidized to form nitric acid are of increasing importance due to stricter controls on emissions of sulfur containing compounds. 70 Tg(S) per year in the form of SO_2 comes from fossil fuel combustion and industry, 2.8 Tg(S) from wildfires and 7–8 Tg(S) per year from volcanoes.

Natural Phenomena

The principal natural phenomena that contribute acid-producing gases to the atmosphere are emissions from volcanoes. Thus, for example, fumaroles from the Laguna Caliente crater of Poás Volcano create extremely high amounts of acid rain and fog, with acidity as high as a pH of 2, clearing an area of any vegetation and frequently causing irritation to the eyes and lungs of inhabitants in nearby settlements. Acid-producing gasses are also created by biological processes that occur on the land, in wetlands, and in the oceans. The major biological source of sulfur containing compounds is dimethyl sulfide.

Nitric acid in rainwater is an important source of fixed nitrogen for plant life, and is also produced by electrical activity in the atmosphere such as lightning.

Acidic deposits have been detected in glacial ice thousands of years old in remote parts of the globe.

Soils of coniferous forests are naturally very acidic due to the shedding of needles, and the results of this phenomenon should not be confused with acid rain.

Human Activity

The coal-fired Gavin Power Plant in Cheshire, Ohio

The principal cause of acid rain is sulfur and nitrogen compounds from human sources, such as electricity generation, factories, and motor vehicles. Electrical power generation using coal is among the greatest contributors to gaseous pollutions that are responsible for acidic rain. The gases can be carried hundreds of kilometers in the atmosphere before they are converted to acids and deposited. In the past, factories had short funnels to let out smoke but this caused many problems locally; thus, factories now have taller smoke funnels. However, dispersal from these taller stacks causes pollutants to be carried farther, causing widespread ecological damage.

Chemical Processes

Combustion of fuels produces sulfur dioxide and nitric oxides. They are converted into sulfuric acid and nitric acid.

Gas Phase Chemistry

In the gas phase sulfur dioxide is oxidized by reaction with the hydroxyl radical via an intermolecular reaction:

$$SO_2 + OH\cdot \rightarrow HOSO_2\cdot$$

which is followed by:

$$HOSO_2\cdot + O_2 \rightarrow HO_2\cdot + SO_3$$

In the presence of water, sulfur trioxide (SO_3) is converted rapidly to sulfuric acid:

$$SO_3\,(g) + H_2O\,(l) \rightarrow H_2SO_4\,(aq)$$

Nitrogen dioxide reacts with OH to form nitric acid:

$NO_2 + OH\cdot \rightarrow HNO_3$

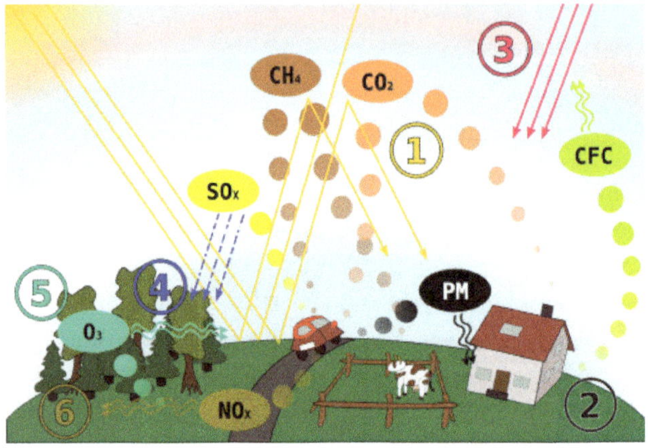

This shows the process of the air pollution being released into the atmosphere and the areas that will be affected.

$$NO_2 + OH\cdot \rightarrow HNO_3$$

Chemistry in Cloud Droplets

When clouds are present, the loss rate of SO_2 is faster than can be explained by gas phase chemistry alone. This is due to reactions in the liquid water droplets.

Hydrolysis

Sulfur dioxide dissolves in water and then, like carbon dioxide, hydrolyses in a series of equilibrium reactions:

$$SO_2\,(g) + H_2O \rightleftharpoons SO_2{\cdot}H_2O$$

$$SO_2{\cdot}H_2O \rightleftharpoons H^+ + HSO_3^-$$

$$HSO_3^- \rightleftharpoons H^+ + SO_3^{2-}$$

Oxidation

There are a large number of aqueous reactions that oxidize sulfur from S(IV) to S(VI), leading to the formation of sulfuric acid. The most important oxidation reactions are with ozone, hydrogen peroxide and oxygen (reactions with oxygen are catalyzed by iron and manganese in the cloud droplets).

Acid Deposition

Wet Deposition

Wet deposition of acids occurs when any form of precipitation (rain, snow, and so on.) removes acids from the atmosphere and delivers it to the Earth's surface. This can result from the deposition of acids produced in the raindrops or by the precipitation removing the acids either in clouds or below clouds. Wet removal of both gases and aerosols are both of importance for wet deposition.

Dry Deposition

Acid deposition also occurs via dry deposition in the absence of precipitation. This can be responsible for as much as 20 to 60% of total acid deposition. This occurs when particles and gases stick to the ground, plants or other surfaces.

Adverse Effects

Acid rain has been shown to have adverse impacts on forests, freshwaters and soils, killing insect and aquatic life-forms as well as causing damage to buildings and having impacts on human health.

Surface Waters and Aquatic Animals

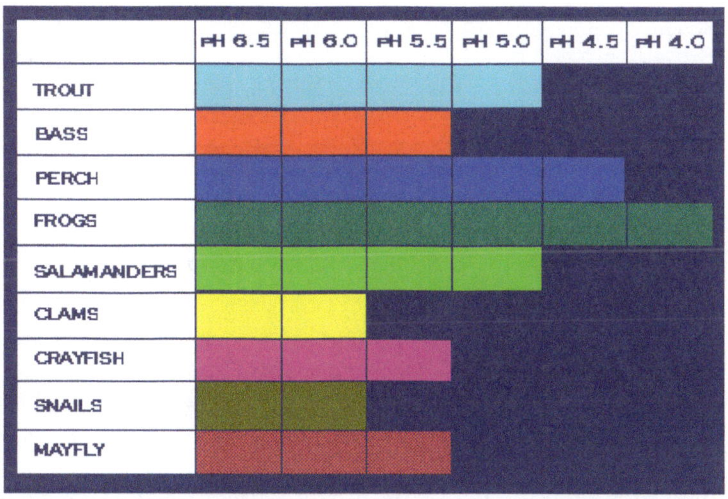

This chart shows that not all fish, shellfish, or the insects that they eat can tolerate the same amount of acid; for example, frogs can tolerate water that is more acidic (i.e., has a lower pH) than trout.

Both the lower pH and higher aluminium concentrations in surface water that occur as a result of acid rain can cause damage to fish and other aquatic animals. At pHs lower than 5 most fish eggs will not hatch and lower pHs can kill adult fish. As lakes and rivers become more acidic biodiversity is reduced. Acid rain has eliminated insect life and some fish species, including the brook trout in some lakes, streams, and creeks in geographically sensitive areas, such as the Adirondack Mountains of the United States. However, the extent to which acid rain contributes directly or indirectly via runoff from the catchment to lake and river acidity (i.e., depending on characteristics of the surrounding watershed) is variable. The United States Environmental Protection Agency's (EPA) website states: "Of the lakes and streams surveyed, acid rain caused acidity in 75% of the acidic lakes and about 50% of the acidic streams".

Soils

Soil biology and chemistry can be seriously damaged by acid rain. Some microbes are unable to tolerate changes to low pH and are killed. The enzymes of these microbes are denatured (changed in shape so they no longer function) by the acid. The hydronium ions of acid rain also mobilize toxins such as aluminium, and leach away essential nutrients and minerals such as magnesium.

$$2\ H^+\ (aq) + Mg^{2+}\ (clay) \rightleftharpoons 2\ H^+\ (clay) + Mg^{2+}\ (aq)$$

Soil chemistry can be dramatically changed when base cations, such as calcium and magnesium, are leached by acid rain thereby affecting sensitive species, such as sugar maple (Acer saccharum).

Forests and Other Vegetation

Adverse effects may be indirectly related to acid rain, like the acid's effects on soil or high concentration of gaseous precursors to acid rain. High altitude forests are especially vulnerable as they are often surrounded by clouds and fog which are more acidic than rain.

Acid rain can have severe effects on vegetation. A forest in the Black Triangle in Europe.

Other plants can also be damaged by acid rain, but the effect on food crops is minimized by the application of lime and fertilizers to replace lost nutrients. In cultivated areas, limestone may also be added to increase the ability of the soil to keep the pH stable, but this tactic is largely unusable in the case of wilderness lands. When calcium is leached from the needles of red spruce, these trees become less cold tolerant and exhibit winter injury and even death.

Ocean Acidification

Acid rain has a much less harmful effect on the oceans. However, acid rain can cause the oceans acidity to rise, making it more difficult for different coastal species to create their exoskeletons that they need to survive. These coastal species link together as part of the ocean's food chain and without them being a source for other marine life to feed off of more marine life will die.

Coral's limestone skeletal is sensitive to pH drop, because the calcium carbonate, core component of the limestone dissolves in acidic (low pH) solutions.

Human Health Effects

Acid rain does not directly affect human health. The acid in the rainwater is too dilute to have direct adverse effects. However, the particulates responsible for acid rain (sulfur dioxide and nitrogen oxides) do have an adverse effect. Increased amounts of fine particulate matter in the air do contribute to heart and lung problems including asthma and bronchitis.

Other Adverse Effects

Effect of acid rain on statues

Acid rain and weathering

Acid rain can damage buildings, historic monuments, and statues, especially those made of rocks, such as limestone and marble, that contain large amounts of calcium carbonate. Acids in the rain react with the calcium compounds in the stones to create gypsum, which then flakes off.

$$CaCO_3\,(s) + H_2SO_4\,(aq) \rightleftharpoons CaSO_4\,(s) + CO_2\,(g) + H_2O\,(l)$$

The effects of this are commonly seen on old gravestones, where acid rain can cause the inscriptions to become completely illegible. Acid rain also increases the corrosion rate of metals, in particular iron, steel, copper and bronze.

Affected Areas

Places significantly impacted by acid rain around the globe include most of eastern Europe from Poland northward into Scandinavia, the eastern third of the United States, and southeastern Canada. Other affected areas include the southeastern coast of China and Taiwan.

Prevention Methods

Technical Solutions

Many coal-firing power stations use flue-gas desulfurization (FGD) to remove sulfur-containing gases from their stack gases. For a typical coal-fired power station, FGD will remove 95% or more of the SO_2 in the flue gases. An example of FGD is the wet scrubber which is commonly used. A wet scrubber is basically a reaction tower equipped with a fan that extracts hot smoke stack gases from a power plant into the tower. Lime or limestone in slurry form is also injected into the tower to mix with the stack gases and combine with the sulfur dioxide present. The calcium carbonate of the limestone produces pH-neutral calcium sulfate that is physically removed from the scrubber. That is, the scrubber turns sulfur pollution into industrial sulfates.

In some areas the sulfates are sold to chemical companies as gypsum when the purity of calcium sulfate is high. In others, they are placed in landfill. However, the effects of acid rain can last for generations, as the effects of pH level change can stimulate the continued leaching of undesirable chemicals into otherwise pristine water sources, killing off vulnerable insect and fish species and blocking efforts to restore native life.

Fluidized bed combustion also reduces the amount of sulfur emitted by power production.

Vehicle emissions control reduces emissions of nitrogen oxides from motor vehicles.

International Treaties

A number of international treaties on the long-range transport of atmospheric pollutants have been agreed for example, the 1985 Helsinki Protocol on the Reduction of Sulphur Emissions under the Convention on Long-Range Transboundary Air Pollution. Canada and the US signed the Air Quality Agreement in 1991. Most European countries and Canada have signed the treaties.

Emissions Trading

In this regulatory scheme, every current polluting facility is given or may purchase on an open market an emissions allowance for each unit of a designated pollutant it emits. Operators can then install pollution control equipment, and sell portions of their emissions allowances they no longer need for their own operations, thereby recovering some of the capital cost of their investment in such equipment. The intention is to give operators economic incentives to install pollution controls.

The first emissions trading market was established in the United States by enactment of the Clean Air Act Amendments of 1990. The overall goal of the Acid Rain Program established by the Act is to achieve significant environmental and public health benefits through reductions in emissions of sulfur dioxide (SO_2) and nitrogen oxides (NO_x), the primary causes of acid rain. To achieve this goal at the lowest cost to society, the program employs both regulatory and market based approaches for controlling air pollution.

Ozone Layer

1. Why the hole in the atmospheric ozone layer above us?
2. What has been done about this?
3. Why should we worry about Ozone Layer Depletion?

Mr. OZONE: Will I be able to continue enjoying this life without problems?

- Ozone (O_3) is a highly-reactive from of oxygen.
- Unlike oxygen (O_2), ozone has a strong scent and is blue in color.
- Ozone exists within both the tropospheric and stratospheric zones of the Earth's atmosphere
- In the troposphere , ground level ozone is a major air pollutant and primary constituent of photochemical smog
- In the stratosphere, the ozone layer is an essential protector of life on earth as it absorbs harmful UV radiation before it reaches the earth.

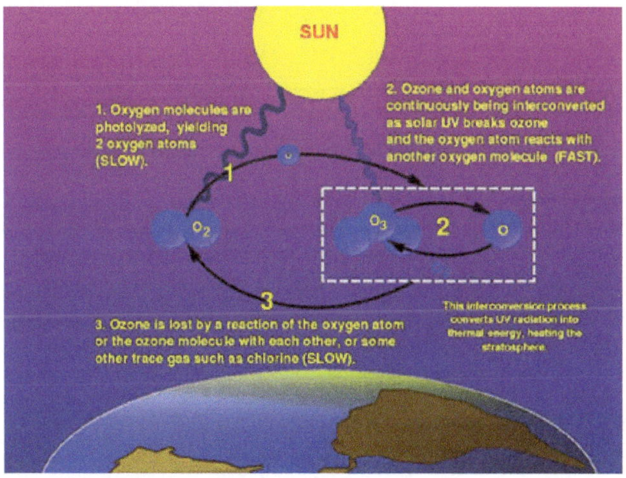

CFC's and Ozone Depletion

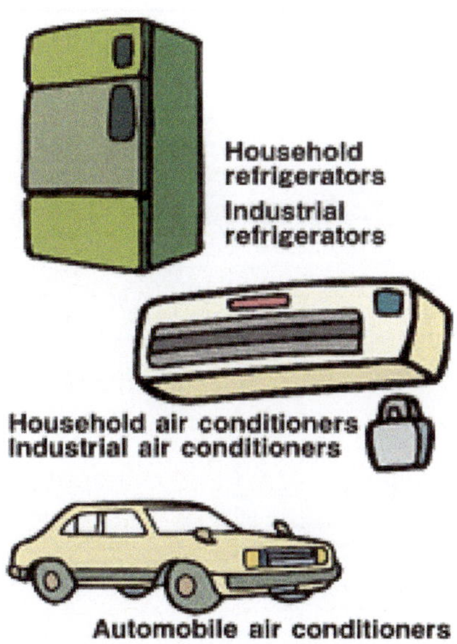

- Chlorofluorocarbons are created and used in refrigerators and air conditioners. These chlorofluorocarbons are not harmful to humans and have been a benefit to us. Once released into the atmosphere, chlorofluorocarbons are bombarded and destroyed by ultraviolet rays. In the process chlorine is released to destroy the ozone molecules.

Destruction of Ozone by Chlorine

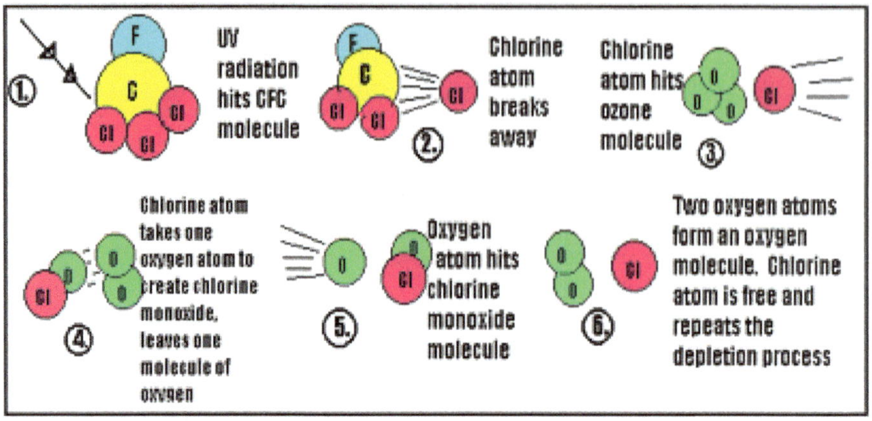

Molecular oxygen is broken down in the stratosphere by solar radiation to yield atomic oxygen, which then combines with molecular oxygen to produce ozone. The ozone is then destroyed by chlorine atoms.

Ozone Destruction by UV Rays

- UV radiation from the sun releases the radicals Cl and ClO.
- Ozone is a highly unstable molecule so it readily donates its extra oxygen molecule to free radical species such as hydrogen, bromine, and chlorine.
- These compound species act as catalysts in the breakdown of ozone molecules.

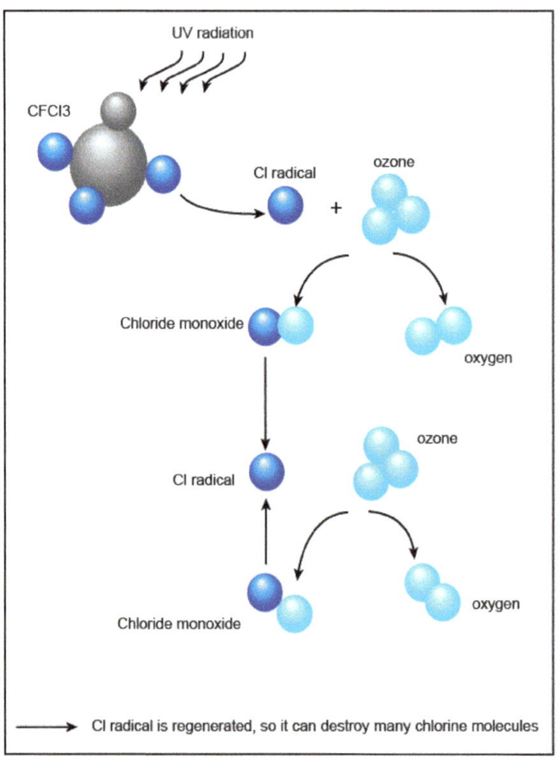

The Ozone Depletion Process

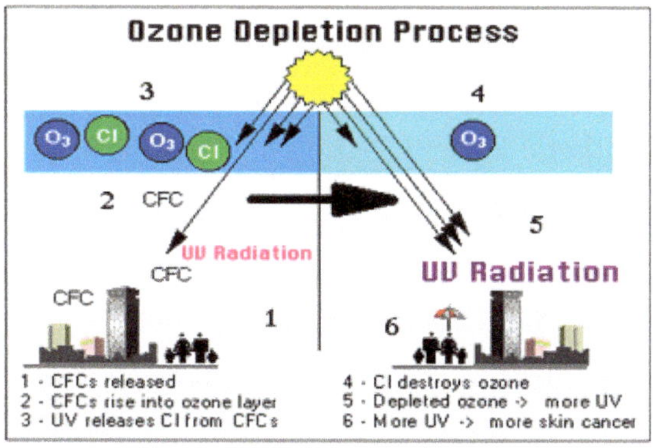

Responsibility for Ozone Damage Each Year

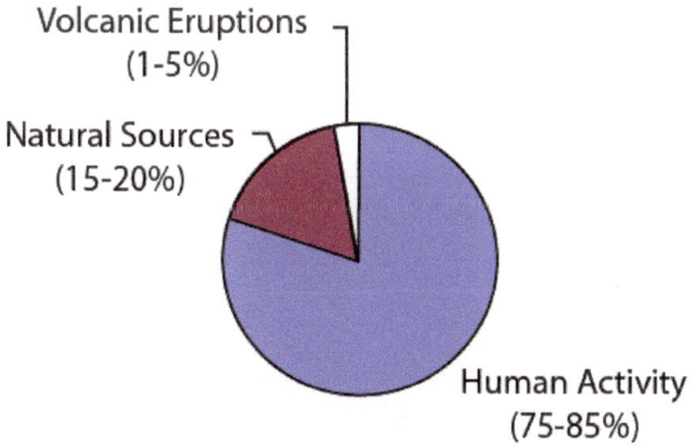

Global CFC Production

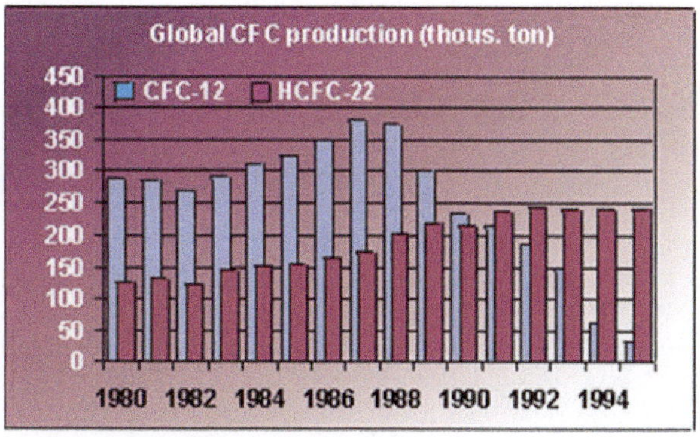

Effects of UV Rays on Aquatic Ecosystems

Ozone depletion causes increases in UV rays's effects on aquatic ecosystems by:

1. decreasing the abundance of phytoplankton – affects the food stock for fishes and the absorption of CO_2
2. decreasing the diversity of aquatic organisms – reduces food stock and also destroys several fish and amphibians.

Effects of UV Rays on Terrestrial Ecosystems

- Damage to plant cell DNA molecules - makes plants more susceptible to pathogens and pests
- Reductions in photosynthetic capacity in the plant - results in slower growth and smaller leaves
- Causes mutations in mammalian cells and destroys membranes

Harmful Effects of UV Rays on People

- Skin cancer
- Premature aging (photoaging) of the skin (different from normal chronological aging)
- Cataracts and eye disorders (corneal sunburn and blindness) Immune system damage

Correlation between Ozone Depletion and Skin Diseases

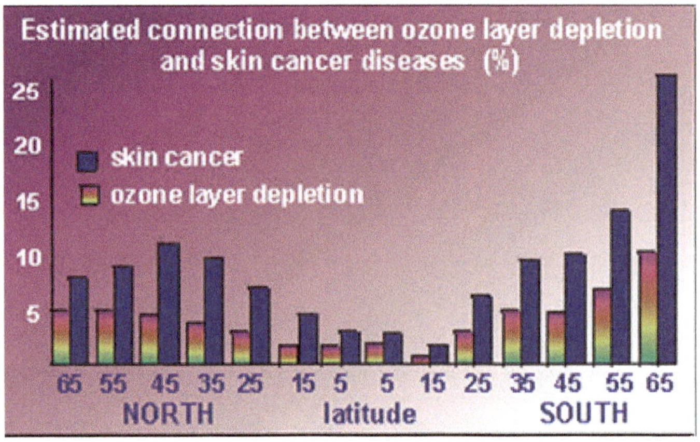

The Ultraviolet Exposure Index

Index Values Exposure

0 – 2 = Minimal

3 - 4 = Low

5 - 6 = Moderate

7 - 9 = High

10 + = Very High

Factors Affecting UV Exposure

- Clouds cover - partly cloudy days do little to reduce UV exposure but rainy or substantially overcast days reduce UV exposures.

- The time of day – peak exposure time is 12:00 noon - 1:00 p.m. UV intensity is reduced by about half at three hours before and three hours after the peak exposure time.

Factors Affecting UV Exposure

- The time of year - more UV is received in the late spring and early summer and much less is received in the late fall and early winter.
- Life style – determines a person's risk to UV exposure. Skiing, sunbathing, or swimming can lead to extremely high exposures. Use of tanning parlors also increase risk.

References

- "The Tropical Atmosphere Ocean Array: Gathering Data to Predict El Niño". Celebrating 200 Years. NOAA. 2007-01-08. Retrieved 2007-07-26
- Berberan-Santos, M. N.; Bodunov, E. N.; Pogliani, L. (1997). "On the barometric formula". American Journal of Physics. 65 (5): 404–412. Bibcode:1997AmJPh..65..404B. doi:10.1119/1.18555
- International Civil Aviation Organization. Manual of the ICAO Standard Atmosphere, Doc 7488-CD, Third Edition, 1993. ISBN 92-9194-004-6
- "Ocean Surface Topography". Oceanography 101. JPL, NASA. 2006-07-05. Archived from the original on April 14, 2009. Retrieved 2007-07-26
- Beychok, Milton R. (1967). Aqueous Wastes from Petroleum and Petrochemical Plants (1st ed.). John Wiley & Sons. ISBN 0-471-07189-7. LCCN 67019834
- "ANNUAL SEA LEVEL DATA SUMMARY REPORT JULY 2005 – JUNE 2006" (pdf). THE AUSTRALIAN BASELINE SEA LEVEL MONITORING PROJECT. Bureau of Meteorology. Retrieved 2007-07-26
- Lackey, R.T. (1997). "Science, policy, and acid rain: lessons learned" (PDF). Renewable Resources Journal. 15 (1): 9–13
- Brian Shmaefsky (2004). Favorite demonstrations for college science: an NSTA Press journals collection. NSTA Press. p. 57. ISBN 978-0-87355-242-4
- David Urbinato (Summer 1994). "London's Historic "Pea-Soupers"". United States Environmental Protection Agency. Retrieved 2006-08-02
- Fleming, James R.; Knorr, Bethany R. "History of the Clean Air Act". American Meteorological Society. Retrieved 2006-02-14
- Dictionary of Chemistry (6 ed.), edited by John Daintith, Publisher: Oxford University Press, 2008, ISBN 9780199204632: "greenhouse effect"
- Waterhealthconnection.org Overview of Waterborne Disease Trends By Patricia L. Meinhardt, MD, MPH, MA, Author. Retrieved on April 16, 2009
- Junling Huang and Michael B. McElroy (2014). "Contributions of the Hadley and Ferrel Circulations to the Energetics of the Atmosphere over the Past 32 Years". Journal of Climate. 27 (7): 2656–2666. Bibcode:2014JCli...27.2656H. doi:10.1175/jcli-d-13-00538.1
- "The Mixture Rule under the Resource Conservation and Recovery Act" (PDF). U.S. Dept. of Energy. 1999. Retrieved 2012-04-10
- Likens, Gene E.; Keene, William C.; Miller, John M.; Galloway, James N. (1987). "Chemistry of precipitation from a remote, terrestrial site in Australia". Journal of Geophysical Research. 92: 13299. Bibcode:1987JGR....9213299L. doi:10.1029/JD092iD11p13299
- Glossary, United States: NASA Earth Observatory, acid rain, archived from the original on 2011-12-13, retrieved 2013-02-15

- Acid Rain in New England, A Brief History Archived September 25, 2010, at the Wayback Machine.. Epa.gov. Retrieved on 2013-02-09
- Keller, C. K.; White, T. M.; O'Brien, R.; Smith, J. L. (2006). "Soil CO2 dynamics and fluxes as affected by tree harvest in an experimental sand ecosystem". Journal of Geophysical Research. 111. Bibcode:2006JGRG..111.3011K. doi:10.1029/2005JG000157
- Chandru (2006-09-09). "CHINA: Industrialization pollutes its country side with Acid Rain". Southasiaanalysis.org. Archived from the original on June 20, 2010. Retrieved 2010-11-18
- Friedman, Sharon M. (2011). "Three Mile Island, Chernobyl, and Fukushima: An analysis of traditional and new media coverage of nuclear accidents and radiation". Bulletin of the atomic scientists. 67 (5): 55–65. doi:10.1177/0096340211421587
- Gerdes, Justin. "Cap and Trade Curbed Acid Rain: 7 Reasons Why It Can Do The Same For Climate Change". Forbes. Forbes. Retrieved 27 Oct 2014
- Zivin, Joshua Graff; Neidell, Matthew (2012-12-01). "The Impact of Pollution on Worker Productivity". American Economic Review. 102 (7): 3652–3673. doi:10.1257/aer.102.7.3652. ISSN 0002-8282. PMC 4576916. PMID 26401055
- Franklin, Ben A. (1984-08-18). "LEGISLATORS SAT WHITE HOUSE SUPPRESSED ACID RAIN REPORT". The New York Times. ISSN 0362-4331. Retrieved 2016-11-16
- Hall, Ronald J.; Likens, Gene E.; Fiance, Sandy B.; Hendrey, George R. (1980-08-01). "Experimental Acidification of a Stream in the Hubbard Brook Experimental Forest, New Hampshire". Ecology. 61 (4): 976–989. ISSN 1939-9170. doi:10.2307/1936765

2

Chemical Pollutants Causing Air Pollution

The presence of harmful substances in the Earth's atmosphere is called air pollution. The chapter delves into air pollution while keeping chemical pollutants in focus. Air pollutants can be categorised into particulate matter, odors, and gaseous pollutants. The section strategically encompasses and incorporates the major components and key concepts of air pollution, providing a complete understanding.

Air Pollution

Air pollution from a coking oven

Air pollution occurs when harmful substances including particulates and biological molecules are introduced into Earth's atmosphere. It may cause diseases, allergies or death in humans; it may also cause harm to other living organisms such as animals and food crops, and may damage the natural or built environment. Human activity and natural processes can both generate air pollution.

Indoor air pollution and poor urban air quality are listed as two of the world's worst toxic pollution problems in the 2008 Blacksmith Institute World's Worst Polluted Places report. According to the 2014 WHO report, air pollution in 2012 caused the deaths of around 7 million people worldwide, an estimate roughly matched by the International Energy Agency.

Pollutants

An air pollutant is a substance in the air that can have adverse effects on humans and the ecosystem. The substance can be solid particles, liquid droplets, or gases. A pollutant can be of natural origin or man-made. Pollutants are classified as primary or secondary. Primary pollutants are usually produced from a process, such as ash from a volcanic eruption. Other examples include carbon monoxide gas from motor vehicle exhaust, or the sulfur dioxide released from factories. Secondary pollutants are not emitted directly. Rather, they form in the air when primary pollutants react or interact. Ground level ozone is a prominent example of a secondary pollutant. Some pollutants may be both primary and secondary: they are both emitted directly and formed from other primary pollutants.

Before flue-gas desulfurization was installed, the emissions from this power plant in New Mexico contained excessive amounts of sulfur dioxide.

Thermal oxidizers are air pollution abatement options for hazardous air pollutants (HAPs), volatile organic compounds (VOCs), and odorous emissions.

Major primary pollutants produced by human activity include:

- Sulfur oxides (SO_x) - particularly sulfur dioxide, a chemical compound with the formula SO_2. SO_2 is produced by volcanoes and in various industrial processes. Coal and petroleum often contain sulfur compounds, and their combustion generates sulfur dioxide. Further oxidation of SO_2, usually in the presence of a catalyst such as NO_2, forms H_2SO_4, and thus acid rain. This is one of the causes for concern over the environmental impact of the use of these fuels as power sources.

- Nitrogen oxides (NO_x) - Nitrogen oxides, particularly nitrogen dioxide, are expelled from high temperature combustion, and are also produced during thunderstorms by electric discharge. They can be seen as a brown haze dome above or a plume downwind of cities. Nitrogen dioxide is a chemical compound with the formula NO_2. It is one of several nitrogen oxides. One of the most prominent air pollutants, this reddish-brown toxic gas has a characteristic sharp, biting odor.

- Carbon monoxide (CO) - CO is a colorless, odorless, toxic yet non-irritating gas. It is a product of incomplete combustion of fuel such as natural gas, coal or wood. Vehicular exhaust is a major source of carbon monoxide.

- Volatile organic compounds (VOC) - VOCs are a well-known outdoor air pollutant. They are categorized as either methane (CH_4) or non-methane (NMVOCs). Methane is an extremely efficient greenhouse gas which contributes to enhanced global warming. Other hydrocarbon VOCs are also significant greenhouse gases because of their role in creating ozone and prolonging the life of methane in the atmosphere. This effect varies depending on local air quality. The aromatic NMVOCs benzene, toluene and xylene are suspected carcinogens and may lead to leukemia with prolonged exposure. 1,3-butadiene is another dangerous compound often associated with industrial use.

- Particulates, alternatively referred to as particulate matter (PM), atmospheric particulate matter, or fine particles, are tiny particles of solid or liquid suspended in a gas. In contrast, aerosol refers to combined particles and gas. Some particulates occur naturally, originating from volcanoes, dust storms, forest and grassland fires, living vegetation, and sea spray. Human activities, such as the burning of fossil fuels in vehicles, power plants and various industrial processes also generate significant amounts of aerosols. Averaged worldwide, anthropogenic aerosols—those made by human activities—currently account for approximately 10 percent of our atmosphere. Increased levels of fine particles in the air are linked to health hazards such as heart disease, altered lung function and lung cancer.

- Persistent free radicals connected to airborne fine particles are linked to cardiopulmonary disease.

- Toxic metals, such as lead and mercury, especially their compounds.

- Chlorofluorocarbons (CFCs) - harmful to the ozone layer; emitted from products are currently banned from use. These are gases which are released from air conditioners, refrigerators, aerosol sprays, etc. On release into the air, CFCs rise to the stratosphere. Here they come in contact with other gases and damage the ozone layer. This allows harmful ultraviolet rays to reach the earth's surface. This can lead to skin cancer, eye disease and can even cause damage to plants.

- Ammonia (NH_3) - emitted from agricultural processes. Ammonia is a compound with the formula NH_3. It is normally encountered as a gas with a characteristic pungent odor. Ammonia contributes significantly to the nutritional needs of terrestrial organisms by serving as a precursor to foodstuffs and fertilizers. Ammonia, either directly or indirectly, is also a building block for the synthesis of many pharmaceuticals. Although in wide use, ammonia is both caustic and hazardous. In the atmosphere, ammonia reacts with oxides of nitrogen and sulfur to form secondary particles.

- Odours — such as from garbage, sewage, and industrial processes

- Radioactive pollutants - produced by nuclear explosions, nuclear events, war explosives, and natural processes such as the radioactive decay of radon.

Secondary pollutants include:

- Particulates created from gaseous primary pollutants and compounds in photochemical smog. Smog is a kind of air pollution. Classic smog results from large amounts of coal burning in an area caused by a mixture of smoke and sulfur dioxide. Modern smog does not usually come from coal but from vehicular and industrial emissions that are acted on in the atmosphere by ultraviolet light from the sun to form secondary pollutants that also combine with the primary emissions to form photochemical smog.
- Ground level ozone (O_3) formed from NO_x and VOCs. Ozone (O_3) is a key constituent of the troposphere. It is also an important constituent of certain regions of the stratosphere commonly known as the Ozone layer. Photochemical and chemical reactions involving it drive many of the chemical processes that occur in the atmosphere by day and by night. At abnormally high concentrations brought about by human activities (largely the combustion of fossil fuel), it is a pollutant, and a constituent of smog.
- Peroxyacetyl nitrate ($C_2H_3NO_5$) - similarly formed from NO_x and VOCs.

Minor air pollutants include:

- A large number of minor hazardous air pollutants. Some of these are regulated in USA under the Clean Air Act and in Europe under the Air Framework Directive
- A variety of persistent organic pollutants, which can attach to particulates

Persistent organic pollutants (POPs) are organic compounds that are resistant to environmental degradation through chemical, biological, and photolytic processes. Because of this, they have been observed to persist in the environment, to be capable of long-range transport, bioaccumulate in human and animal tissue, biomagnify in food chains, and to have potentially significant impacts on human health and the environment.

Sources

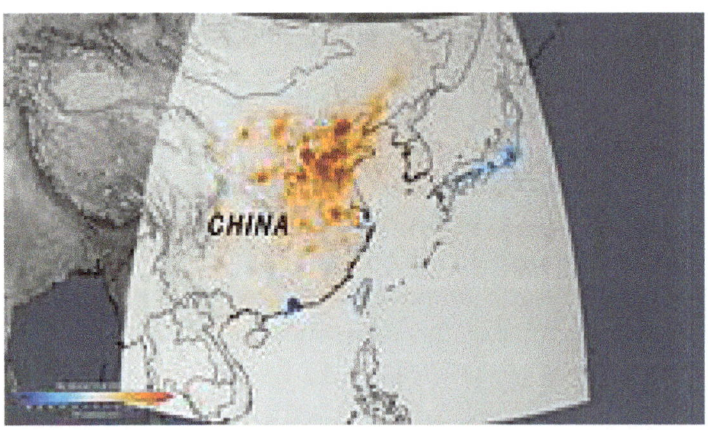

This video provides an overview of a NASA study on the human fingerprint on global air quality.

There are various locations, activities or factors which are responsible for releasing pollutants into the atmosphere. These sources can be classified into two major categories.

Anthropogenic (Man-made) Sources:

Controlled burning of a field outside of Statesboro, Georgia in preparation for spring planting.

These are mostly related to the burning of multiple types of fuel.

- Stationary sources include smoke stacks of power plants, manufacturing facilities (factories) and waste incinerators, as well as furnaces and other types of fuel-burning heating devices. In developing and poor countries, traditional biomass burning is the major source of air pollutants; traditional biomass includes wood, crop waste and dung.
- Mobile sources include motor vehicles, marine vessels, and aircraft.
- Controlled burn practices in agriculture and forest management. Controlled or prescribed burning is a technique sometimes used in forest management, farming, prairie restoration or greenhouse gas abatement. Fire is a natural part of both forest and grassland ecology and controlled fire can be a tool for foresters. Controlled burning stimulates the germination of some desirable forest trees, thus renewing the forest.
- Fumes from paint, hair spray, varnish, aerosol sprays and other solvents
- Waste deposition in landfills, which generate methane. Methane is highly flammable and may form explosive mixtures with air. Methane is also an asphyxiant and may displace oxygen in an enclosed space. Asphyxia or suffocation may result if the oxygen concentration is reduced to below 19.5% by displacement.
- Military resources, such as nuclear weapons, toxic gases, germ warfare and rocketry

Natural sources:

Dust storm approaching Stratford, Texas.

- Dust from natural sources, usually large areas of land with little or no vegetation
- Methane, emitted by the digestion of food by animals, for example cattle
- Radon gas from radioactive decay within the Earth's crust. Radon is a colorless, odorless, naturally occurring, radioactive noble gas that is formed from the decay of radium. It is considered to be a health hazard. Radon gas from natural sources can accumulate in buildings, especially in confined areas such as the basement and it is the second most frequent cause of lung cancer, after cigarette smoking
- Smoke and carbon monoxide from wildfires
- Vegetation, in some regions, emits environmentally significant amounts of Volatile organic compounds (VOCs) on warmer days. These VOCs react with primary anthropogenic pollutants—specifically, NO_x, SO_2, and anthropogenic organic carbon compounds — to produce a seasonal haze of secondary pollutants. Black gum, poplar, oak and willow are some examples of vegetation that can produce abundant VOCs. The VOC production from these species result in ozone levels up to eight times higher than the low-impact tree species.
- Volcanic activity, which produces sulfur, chlorine, and ash particulates.

Emission Factors

Air pollutant emission factors are reported representative values that attempt to relate the quantity of a pollutant released to the ambient air with an activity associated with the release of that pollutant. These factors are usually expressed as the weight of pollutant divided by a unit weight, volume, distance, or duration of the activity emitting the pollutant (e.g., kilograms of particulate emitted per tonne of coal burned). Such factors facilitate estimation of emissions from various sources of air pollution. In most cases, these factors are simply averages of all available data of acceptable quality, and are generally assumed to be representative of long-term averages.

There are 12 compounds in the list of persistent organic pollutants. Dioxins and furans are two of them and intentionally created by combustion of organics, like open burning of plastics. These compounds are also endocrine disruptors and can mutate the human genes.

The United States Environmental Protection Agency has published a compilation of air pollutant emission factors for a wide range of industrial sources. The United Kingdom, Australia, Canada and many other countries have published similar compilations, as well as the European Environment Agency.

Exposure

Air pollution risk is a function of the hazard of the pollutant and the exposure to that pollutant. Air pollution exposure can be expressed for an individual, for certain groups (e.g. neighborhoods or children living in a country), or for entire populations. For example, one may want to calculate the exposure to a hazardous air pollutant for a geographic area, which includes the various microenvironments and age groups. This can be calculated as an inhalation exposure. This would account for daily exposure in various settings (e.g. different indoor micro-environments and outdoor locations). The exposure needs to include different age and other demographic groups, especially infants, children, pregnant women and other sensitive subpopulations. The exposure to an air pollutant must integrate the concentrations of the air pollutant with respect to the time spent in each setting and the respective inhalation rates for each subgroup for each specific time that the subgroup is in the setting and engaged in particular activities (playing, cooking, reading, working, etc.). For example, a small child's inhalation rate will be less than that of an adult. A child engaged in vigorous exercise will have a higher respiration rate than the same child in a sedentary activity. The daily exposure, then, needs to reflect the time spent in each micro-environmental setting and the type of activities in these settings. The air pollutant concentration in each microactivity/micro-environmental setting is summed to indicate the exposure.

Indoor Air Quality (IAQ)

Air quality monitoring, New Delhi, India.

A lack of ventilation indoors concentrates air pollution where people often spend the majority of their time. Radon (Rn) gas, a carcinogen, is exuded from the Earth in certain locations and trapped inside houses. Building materials including carpeting and plywood emit formaldehyde (H_2CO) gas. Paint and solvents give off volatile organic compounds (VOCs) as they dry. Lead paint can degenerate into dust and be inhaled. Intentional air pollution is introduced with the use of air fresheners, incense, and other scented items. Controlled wood fires in stoves and fireplaces can add significant amounts of smoke particulates into the air, inside and out. Indoor pollution fatalities may be caused by using pesticides and other chemical sprays indoors without proper ventilation.

Carbon monoxide poisoning and fatalities are often caused by faulty vents and chimneys, or by the burning of charcoal indoors or in a confined space, such as a tent. Chronic carbon monoxide poisoning can result even from poorly-adjusted pilot lights. Traps are built into all domestic plumbing to keep sewer gas and hydrogen sulfide, out of interiors. Clothing emits tetrachloroethylene, or other dry cleaning fluids, for days after dry cleaning.

Though its use has now been banned in many countries, the extensive use of asbestos in industrial and domestic environments in the past has left a potentially very dangerous material in many localities. Asbestosis is a chronic inflammatory medical condition affecting the tissue of the lungs. It occurs after long-term, heavy exposure to asbestos from asbestos-containing materials in structures. Sufferers have severe dyspnea (shortness of breath) and are at an increased risk regarding several different types of lung cancer. As clear explanations are not always stressed in non-technical literature, care should be taken to distinguish between several forms of relevant diseases.

According to the World Health Organisation (WHO), these may defined as; asbestosis, *lung cancer*, and *Peritoneal Mesothelioma* (generally a very rare form of cancer, when more widespread it is almost always associated with prolonged exposure to asbestos).

Biological sources of air pollution are also found indoors, as gases and airborne particulates. Pets produce dander, people produce dust from minute skin flakes and decomposed hair, dust mites in bedding, carpeting and furniture produce enzymes and micrometre-sized fecal droppings, inhabitants emit methane, mold forms on walls and generates mycotoxins and spores, air conditioning systems can incubate Legionnaires' disease and mold, and houseplants, soil and surrounding gardens can produce pollen, dust, and mold. Indoors, the lack of air circulation allows these airborne pollutants to accumulate more than they would otherwise occur in nature.

Health Effects

Air pollution is a significant risk factor for a number of pollution-related diseases and health conditions including respiratory infections, heart disease, COPD, stroke and lung cancer. The health effects caused by air pollution may include difficulty in breathing, wheezing, coughing, asthma and worsening of existing respiratory and cardiac conditions. These effects can result in increased medication use, increased doctor or emergency room visits, more hospital admissions and premature death. The human health effects of poor air quality are far reaching, but principally affect the body's respiratory system and the cardiovascular system. Individual reactions to air pollutants depend on the type of pollutant a person is exposed to, the degree of exposure, and the individual's health status and genetics. The most common sources of air pollution include particulates, ozone, nitrogen dioxide, and sulphur dioxide. Children aged less than five years that live in developing countries are the most vulnerable population in terms of total deaths attributable to indoor and outdoor air pollution.

Mortality

The World Health Organization estimated in 2014 that every year air pollution causes the premature death of some 7 million people worldwide. India has the highest death rate due to air pollution. India also has more deaths from asthma than any other nation according to the World Health Organization. In December 2013 air pollution was estimated to kill 500,000 people in China each year. There is a positive correlation between pneumonia-related deaths and air pollution from motor vehicle emissions.

Annual premature European deaths caused by air pollution are estimated at 430,000. An important cause of these deaths is nitrogen dioxide and other nitrogen oxides (NOx) emitted by road vehicles. Across the European Union, air pollution is estimated to reduce life expectancy by almost nine months. Causes of deaths include strokes, heart disease, COPD, lung cancer, and lung infections.

Urban outdoor air pollution is estimated to cause 1.3 million deaths worldwide per year. Children are particularly at risk due to the immaturity of their respiratory organ systems.

The US EPA estimates that a proposed set of changes in diesel engine technology (*Tier 2*) could result in 12,000 fewer *premature mortalities*, 15,000 fewer heart attacks, 6,000 fewer emergency room visits by children with asthma, and 8,900 fewer respiratory-related hospital admissions each year in the United States.

The US EPA has estimated that limiting ground-level ozone concentration to 65 parts per billion, would avert 1,700 to 5,100 premature deaths nationwide in 2020 compared with the 75-ppb standard. The agency projected the more protective standard would also prevent an additional 26,000 cases of aggravated asthma, and more than a million cases of missed work or school. Following this assessment, the EPA acted to protect public health by lowering the National Ambient Air Quality Standards (NAAQS) for ground-level ozone to 70 parts per billion (ppb).

A new economic study of the health impacts and associated costs of air pollution in the Los Angeles Basin and San Joaquin Valley of Southern California shows that more than 3,800 people die prematurely (approximately 14 years earlier than normal) each year because air pollution levels violate federal standards. The number of annual premature deaths is considerably higher than the fatalities related to auto collisions in the same area, which average fewer than 2,000 per year.

Diesel exhaust (DE) is a major contributor to combustion-derived particulate matter air pollution. In several human experimental studies, using a well-validated exposure chamber setup, DE has been linked to acute vascular dysfunction and increased thrombus formation.

The mechanisms linking air pollution to increased cardiovascular mortality are uncertain, but probably include pulmonary and systemic inflammation.

Cardiovascular Disease

A 2007 review of evidence found ambient air pollution exposure is a risk factor correlating with increased total mortality from cardiovascular events (range: 12% to 14% per 10 microg/m^3 increase).

Air pollution is also emerging as a risk factor for stroke, particularly in developing countries where pollutant levels are highest. A 2007 study found that in women, air pollution is not associated with hemorrhagic but with ischemic stroke. Air pollution was also found to be associated with increased incidence and mortality from coronary stroke in a cohort study in 2011. Associations are believed to be causal and effects may be mediated by vasoconstriction, low-grade inflammation and atherosclerosis Other mechanisms such as autonomic nervous system imbalance have also been suggested.

Lung Disease

Chronic obstructive pulmonary disease (COPD) includes diseases such as chronic bronchitis and emphysema.

Research has demonstrated increased risk of developing asthma and COPD from increased exposure to traffic-related air pollution. Additionally, air pollution has been associated with increased hospitalization and mortality from asthma and COPD.

A study conducted in 1960-1961 in the wake of the Great Smog of 1952 compared 293 London residents with 477 residents of Gloucester, Peterborough, and Norwich, three towns with low reported death rates from chronic bronchitis. All subjects were male postal truck drivers aged 40 to

59. Compared to the subjects from the outlying towns, the London subjects exhibited more severe respiratory symptoms (including cough, phlegm, and dyspnea), reduced lung function (FEV_1 and peak flow rate), and increased sputum production and purulence. The differences were more pronounced for subjects aged 50 to 59. The study controlled for age and smoking habits, so concluded that air pollution was the most likely cause of the observed differences.

It is believed that much like cystic fibrosis, by living in a more urban environment serious health hazards become more apparent. Studies have shown that in urban areas patients suffer mucus hypersecretion, lower levels of lung function, and more self-diagnosis of chronic bronchitis and emphysema.

Cancer

Cancer mainly the result of environmental factors.

A review of evidence regarding whether ambient air pollution exposure is a risk factor for cancer in 2007 found solid data to conclude that long-term exposure to PM2.5 (fine particulates) increases the overall risk of non-accidental mortality by 6% per a 10 microg/m^3 increase. Exposure to PM2.5 was also associated with an increased risk of mortality from lung cancer (range: 15% to 21% per 10 microg/m^3 increase) and total cardiovascular mortality (range: 12% to 14% per a 10 microg/m^3 increase). The review further noted that living close to busy traffic appears to be associated with elevated risks of these three outcomes --- increase in lung cancer deaths, cardiovascular deaths, and overall non-accidental deaths. The reviewers also found suggestive evidence that exposure to PM2.5 is positively associated with mortality from coronary heart diseases and exposure to SO_2 increases mortality from lung cancer, but the data was insufficient to provide solid conclusions. Another investigation showed that higher activity level increases deposition fraction of aerosol particles in human lung and recommended avoiding heavy activities like running in outdoor space at polluted areas.

In 2011, a large Danish epidemiological study found an increased risk of lung cancer for patients who lived in areas with high nitrogen oxide concentrations. In this study, the association was higher for non-smokers than smokers. An additional Danish study, also in 2011, likewise noted evidence of possible associations between air pollution and other forms of cancer, including cervical cancer and brain cancer.

In December 2015, medical scientists reported that cancer is overwhelmingly a result of environmental factors, and not largely down to bad luck. Maintaining a healthy weight, eating a healthy diet, minimizing alcohol and eliminating smoking reduces the risk of developing the disease, according to the researchers.

Children

In the United States, despite the passage of the Clean Air Act in 1970, in 2002 at least 146 million Americans were living in non-attainment areas—regions in which the concentration of certain air pollutants exceeded federal standards. These dangerous pollutants are known as the criteria pollutants, and include ozone, particulate matter, sulfur dioxide, nitrogen dioxide, carbon monoxide, and lead. Protective measures to ensure children's health are being taken in cities such as New Delhi, India where buses now use compressed natural gas to help eliminate the "pea-soup" smog. A recent study in Europe has found that exposure to ultrafine particles can increase blood pressure in children.

Infants

Ambient levels of air pollution have been associated with preterm birth and low birth weight. A 2014 WHO worldwide survey on maternal and perinatal health found a statistically significant association between low birth weights (LBW) and increased levels of exposure to PM2.5. Women in regions with greater than average PM2.5 levels had statistically significant higher odds of pregnancy resulting in a low-birth weight infant even when adjusted for country-related variables. The effect is thought to be from stimulating inflammation and increasing oxidative stress.

A study by the University of York found that in 2010 exposure to PM2.5 was strongly associated with 18% of preterm births globally, which was approximately 2.7 million premature births. The countries with the highest air pollution associated preterm births were in South and East Asia, the Middle East, North Africa, and West sub-Saharan Africa.

The source of PM 2.5 differs greatly by region. In South and East Asia, pregnant women are frequently exposed to indoor air pollution because of the wood and other biomass fuels used for cooking which are responsible for more than 80% of regional pollution. In the Middle East, North Africa and West sub-Saharan Africa, fine PM comes from natural sources, such as dust storms. The United States had an estimated 50,000 preterm births associated with exposure to PM2.5 in 2010.

A study performed by Wang, et al. between the years of 1988 and 1991 has found a correlation between Sulfur Dioxide (SO2) and total suspended particulates (TSP) and preterm births and low birth weights in Beijing. A group of 74,671 pregnant women, in four separate regions of Beijing, were monitored from early pregnancy to delivery along with daily air pollution levels of Sulfur Dioxide and TSP (along with other particulates). The estimated reduction in birth weight was 7.3 g for every 100 μg/m3 increase in SO2 and 6.9g for each 100 μg/m3 increase in TSP. These associations were statistically significant in both summer and winter, although, summer was greater. The proportion of low birth weight attributable to air pollution, was 13%. This is the largest attributable risk ever reported for the known risk factors of low birth weight. Coal stoves, which are in 97% of homes, are a major source of air pollution in this area.

Brauer et al studied the relationship between air pollution and proximity to a highway with pregnancy outcomes in a Vancouver cohort of pregnant woman using addresses to estimate exposure during pregnancy. Exposure to NO, NO2, CO PM10 and PM2.5 were associated with infants born small for gestational age (SGA). Women living <50meters away from an expressway or highway were 26% more likely to give birth to a SGA infant.

Clean Areas

Even in the areas with relatively low levels of air pollution, public health effects can be significant and costly, since a large number of people breathe in such pollutants. A 2005 scientific study for the British Columbia Lung Association showed that a small improvement in air quality (1% reduction of ambient PM2.5 and ozone concentrations) would produce $29 million in annual savings in the Metro Vancouver region in 2010. This finding is based on health valuation of lethal (death) and sub-lethal (illness) affects.

Central Nervous System

Data is accumulating that air pollution exposure also affects the central nervous system.

In a June 2014 study conducted by researchers at the University of Rochester Medical Center, published in the journal Environmental Health Perspectives, it was discovered that early exposure to air pollution causes the same damaging changes in the brain as autism and schizophrenia. The study also shows that air pollution also affected short-term memory, learning ability, and impulsivity. Lead researcher Professor Deborah Cory-Slechta said that "When we looked closely at the ventricles, we could see that the white matter that normally surrounds them hadn't fully developed. It appears that inflammation had damaged those brain cells and prevented that region of the brain from developing, and the ventricles simply expanded to fill the space. Our findings add to the growing body of evidence that air pollution may play a role in autism, as well as in other neurodevelopmental disorders." Air pollution has a more significant negative effect on males than on females.

In 2015, experimental studies reported the detection of significant episodic (situational) cognitive impairment from impurities in indoor air breathed by test subjects who were not informed about changes in the air quality. Researchers at the Harvard University and SUNY Upstate Medical University and Syracuse University measured the cognitive performance of 24 participants in three different controlled laboratory atmospheres that simulated those found in "conventional" and "green" buildings, as well as green buildings with enhanced ventilation. Performance was evaluated objectively using the widely used Strategic Management Simulation software simulation tool, which is a well-validated assessment test for executive decision-making in an unconstrained situation allowing initiative and improvisation. Significant deficits were observed in the performance scores achieved in increasing concentrations of either volatile organic compounds (VOCs) or carbon dioxide, while keeping other factors constant. The highest impurity levels reached are not uncommon in some classroom or office environments.

Agricultural Effects

In India in 2014, it was reported that air pollution by black carbon and ground level ozone had cut crop yields in the most affected areas by almost half in 2010 when compared to 1980 levels.

Economic Effects

Air pollution costs the world economy $5 trillion per year as a result of productivity losses and degraded quality of life, according to a joint study by the World Bank and the Institute for Health Metrics and Evaluation (IHME) at the University of Washington These productivity losses are caused by deaths due to diseases caused by air pollution. One out of ten deaths in 2013 was caused by diseases associated with air pollution and the problem is getting worse. The problem is even more acute in the developing world. "Children under age 5 in lower-income countries are more than 60 times as likely to die from exposure to air pollution as children in high-income countries." The report states that additional economic losses caused by air pollution, including health costs and the adverse effect on agricultural and other productivity were not calculated in the report, and thus the actual costs to the world economy are far higher than $5 trillion.

Historical Disasters

The world's worst short-term civilian pollution crisis was the 1984 Bhopal Disaster in India. Leaked industrial vapours from the Union Carbide factory, belonging to Union Carbide, Inc., U.S.A. (later bought by Dow Chemical Company), killed at least 3787 people and injured from 150,000 to 600,000. The United Kingdom suffered its worst air pollution event when the December 4 Great Smog of 1952 formed over London. In six days more than 4,000 died and more recent estimates put the figure at nearer 12,000. An accidental leak of anthrax spores from a biological warfare laboratory in the former USSR in 1979 near Sverdlovsk is believed to have caused at least 64 deaths. The worst single incident of air pollution to occur in the US occurred in Donora, Pennsylvania in late October, 1948, when 20 people died and over 7,000 were injured.

Alternatives to Pollution

There are now practical alternatives to the three principal causes of air pollution.

- Combustion of fossil fuels for space heating can be replaced by using ground source heat pumps and seasonal thermal energy storage.
- Electric power generation from burning fossil fuels can be replaced by power generation from nuclear and renewables.
- Motor vehicles driven by fossil fuels, a key factor in urban air pollution, can be replaced by electric vehicles.

Reduction Efforts

There are various air pollution control technologies and strategies available to reduce air pollution. At its most basic level, land-use planning is likely to involve zoning and transport infrastructure planning. In most developed countries, land-use planning is an important part of social policy, ensuring that land is used efficiently for the benefit of the wider economy and population, as well as to protect the environment.

Because a large share of air pollution is caused by combustion of fossil fuels such as coal and oil, the reduction of these fuels can reduce air pollution drastically. Most effective is the switch to clean power sources such as wind power, solar power, hydro power which don't cause air pollution. Efforts to reduce pollution from mobile sources includes primary regulation (many developing countries have permissive regulations), expanding regulation to new sources (such as cruise and transport ships, farm equipment, and small gas-powered equipment such as string trimmers, chainsaws, and snowmobiles), increased fuel efficiency (such as through the use of hybrid vehicles), conversion to cleaner fuels or conversion to electric vehicles.

Titanium dioxide has been researched for its ability to reduce air pollution. Ultraviolet light will release free electrons from material, thereby creating free radicals, which break up VOCs and NOx gases. One form is superhydrophilic.

In 2014, Prof. Tony Ryan and Prof. Simon Armitage of University of Sheffield prepared a 10 meter by 20 meter-sized poster coated with microscopic, pollution-eating nanoparticles of titanium dioxide. Placed on a building, this giant poster can absorb the toxic emission from around 20 cars each day.

A very effective means to reduce air pollution is the transition to renewable energy. According to a study published in Energy and Environmental Science in 2015 the switch to 100% renewable energy in the United States would eliminate about 62,000 premature mortalities per year and about 42,000 in 2050, if no biomass were used. This would save about $600 billion in health costs a year due to reduced air pollution in 2050, or about 3.6% of the 2014 U.S. gross domestic product.

Control Devices

The following items are commonly used as pollution control devices in industry and transportation. They can either destroy contaminants or remove them from an exhaust stream before it is emitted into the atmosphere.

- Particulate control
  - Mechanical collectors (dust cyclones, multicyclones)
  - Electrostatic precipitators An electrostatic precipitator (ESP), or electrostatic air cleaner is a particulate collection device that removes particles from a flowing gas (such as air), using the force of an induced electrostatic charge. Electrostatic precipitators are highly efficient filtration devices that minimally impede the flow of gases through the device, and can easily remove fine particulates such as dust and smoke from the air stream.
  - Baghouses Designed to handle heavy dust loads, a dust collector consists of a blower, dust filter, a filter-cleaning system, and a dust receptacle or dust removal system (distinguished from air cleaners which utilize disposable filters to remove the dust).
  - Particulate scrubbers Wet scrubber is a form of pollution control technology. The term describes a variety of devices that use pollutants from a furnace flue gas or from other gas streams. In a wet scrubber, the polluted gas stream is brought into contact with the scrubbing liquid, by spraying it with the liquid, by forcing it through a pool of liquid, or by some other contact method, so as to remove the pollutants.

- Scrubbers
    - Baffle spray scrubber
    - Cyclonic spray scrubber
    - Ejector venturi scrubber
    - Mechanically aided scrubber
    - Spray tower
    - Wet scrubber
- NOx control
    - Low NOx burners
    - Selective catalytic reduction (SCR)
    - Selective non-catalytic reduction (SNCR)
    - NOx scrubbers
    - Exhaust gas recirculation
    - Catalytic converter (also for VOC control)
- VOC abatement
    - Adsorption systems, using activated carbon, such as Fluidized Bed Concentrator
    - Flares
    - Thermal oxidizers
    - Catalytic converters
    - Biofilters
    - Absorption (scrubbing)
    - Cryogenic condensers
    - Vapor recovery systems
- Acid Gas/SO_2 control
    - Wet scrubbers
    - Dry scrubbers
    - Flue-gas desulfurization
- Mercury control
    - Sorbent Injection Technology
    - Electro-Catalytic Oxidation (ECO)
    - K-Fuel

- Dioxin and furan control
- Miscellaneous associated equipment
    - o Source capturing systems
    - o Continuous emissions monitoring systems (CEMS)

Regulations

Smog in Cairo

In general, there are two types of air quality standards. The first class of standards (such as the U.S. National Ambient Air Quality Standards and E.U. Air Quality Directive) set maximum atmospheric concentrations for specific pollutants. Environmental agencies enact regulations which are intended to result in attainment of these target levels. The second class (such as the North American Air Quality Index) take the form of a scale with various thresholds, which is used to communicate to the public the relative risk of outdoor activity. The scale may or may not distinguish between different pollutants.

Canada

In Canada, air pollution and associated health risks are measured with the Air Quality Health Index or (AQHI). It is a health protection tool used to make decisions to reduce short-term exposure to air pollution by adjusting activity levels during increased levels of air pollution.

The Air Quality Health Index or "AQHI" is a federal program jointly coordinated by Health Canada and Environment Canada. However, the AQHI program would not be possible without the commitment and support of the provinces, municipalities and NGOs. From air quality monitoring to health risk communication and community engagement, local partners are responsible for the vast majority of work related to AQHI implementation. The AQHI provides a number from 1 to 10+ to indicate the level of health risk associated with local air quality. Occasionally, when the amount of air pollution is abnormally high, the number may exceed 10. The AQHI provides a local air quality current value as well as a local air quality maximums forecast for today, tonight and tomorrow and provides associated health advice.

Risk: Low **(1-3)** Moderate **(4-6)** High **(7-10)** Very high **(above 10)**

As it is now known that even low levels of air pollution can trigger discomfort for the sensitive population, the index has been developed as a continuum: The higher the number, the greater the health risk and need to take precautions. The index describes the level of health risk associated with this number as 'low', 'moderate', 'high' or 'very high', and suggests steps that can be taken to reduce exposure.

Health Risk	Air Quality Health Index	Health Messages	
		At Risk population	General Population
Low	1-3	**Enjoy** your usual outdoor activities.	**Ideal** air quality for outdoor activities
Moderate	4-6	**Consider reducing** or rescheduling strenuous activities outdoors if you are experiencing symptoms.	**No need to modify** your usual outdoor activities unless you experience symptoms such as coughing and throat irritation.
High	7-10	**Reduce** or reschedule strenuous activities outdoors. Children and the elderly should also take it easy.	**Consider reducing** or rescheduling strenuous activities outdoors if you experience symptoms such as coughing and throat irritation.
Very high	Above 10	**Avoid** strenuous activities outdoors. Children and the elderly should also avoid outdoor physical exertion and should stay indoors.	**Reduce** or reschedule strenuous activities outdoors, especially if you experience symptoms such as coughing and throat irritation.

The measurement is based on the observed relationship of Nitrogen Dioxide (NO_2), ground-level Ozone (O_3) and particulates ($PM_{2.5}$) with mortality, from an analysis of several Canadian cities. Significantly, all three of these pollutants can pose health risks, even at low levels of exposure, especially among those with pre-existing health problems.

When developing the AQHI, Health Canada's original analysis of health effects included five major air pollutants: particulates, ozone, and nitrogen dioxide (NO2), as well as sulfur dioxide (SO_2), and carbon monoxide (CO). The latter two pollutants provided little information in predicting health effects and were removed from the AQHI formulation.

The AQHI does not measure the effects of odour, pollen, dust, heat or humidity.

Germany

TA Luft is the German air quality regulation.

Hotspots

Air pollution hotspots are areas where air pollution emissions expose individuals to increased negative health effects. They are particularly common in highly populated, urban areas, where there may be a combination of stationary sources (e.g. industrial facilities) and mobile sources (e.g. cars and trucks) of pollution. Emissions from these sources can cause respiratory disease, childhood asthma, cancer, and other health problems. Fine particulate matter such as diesel soot, which contributes to more than 3.2 million premature deaths around the world each year, is a significant problem. It is very small and can lodge itself within the lungs and enter the bloodstream. Diesel soot is concentrated in densely populated areas, and one in six people in the U.S. live near a diesel pollution hot spot.

While air pollution hotspots affect a variety of populations, some groups are more likely to be located in hotspots. Previous studies have shown disparities in exposure to pollution by race and/or income. Hazardous land uses (toxic storage and disposal facilities, manufacturing facilities, major roadways) tend to be located where property values and income levels are low. Low socioeconomic status can be a proxy for other kinds of social vulnerability, including race, a lack of ability to influence regulation and a lack of ability to move to neighborhoods with less environmental pollution. These communities bear a disproportionate burden of environmental pollution and are more likely to face health risks such as cancer or asthma.

Studies show that patterns in race and income disparities not only indicate a higher exposure to pollution but also higher risk of adverse health outcomes. Communities characterized by low socioeconomic status and racial minorities can be more vulnerable to cumulative adverse health impacts resulting from elevated exposure to pollutants than more privileged communities. Blacks and Latinos generally face more pollution than whites and Asians, and low-income communities bear a higher burden of risk than affluent ones. Racial discrepancies are particularly distinct in suburban areas of the US South and metropolitan areas of the US West. Residents in public housing, who are generally low-income and cannot move to healthier neighborhoods, are highly affected by nearby refineries and chemical plants.

Cities

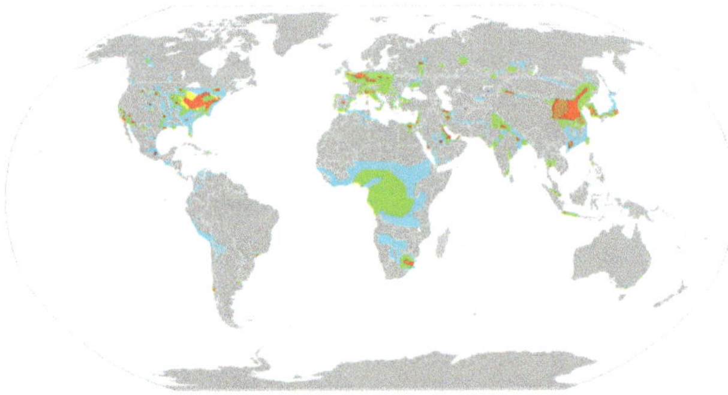

Nitrogen dioxide concentrations as measured from satellite 2002-2004

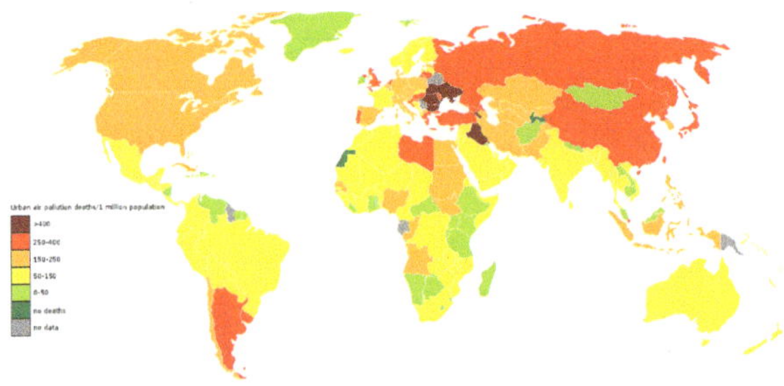

Deaths from air pollution in 2004

Air pollution is usually concentrated in densely populated metropolitan areas, especially in developing countries where environmental regulations are relatively lax or nonexistent. However, even populated areas in developed countries attain unhealthy levels of pollution, with Los Angeles and Rome being two examples. Between 2002 and 2011 the incidence of lung cancer in Beijing near doubled. While smoking remains the leading cause of lung cancer in China, the number of smokers is falling while lung cancer rates are rising. Another project focusing on the effects on pollution in vegetation has been researched by the local university in Sheffield, UK.

National-scale Air Toxics Assessments 1995-2005

The national-scale air toxics assessment(NATA) is an evaluation of air toxics by the U.S. EPA. EPA has furnished four assessments that characterize nationwide chronic cancer risk estimates and noncancer hazards from inhaling air toxics. The lates was from 2005, and made publicly available in early 2011.

"EPA developed the NATA as a state-of-the-science screening tool for State/Local/Tribal Agencies to prioritize pollutants, emission sources and locations of interest for further study, in order to gain a better understanding of the risks. NATA assessments do not incorporate refined information about emission sources, but rather, use general information about sources to develop estimates of risks which are more likely to overestimate impacts than underestimate them. NATA provides estimates of the risk of cancer and other serious health effects from breathing (inhaling) air toxics in order to inform both national and more localized efforts to identify and prioritize air toxics, emission source types and locations which are of greatest potential concern in terms of contributing to population risk. This in turn helps air pollution experts focus limited analytical resources on areas and or populations where the potential for health risks are highest. Assessments include estimates of cancer and non-cancer health effects based on chronic exposure from outdoor sources, including assessments of non-cancer health effects for Diesel Particulate Matter. Assessments provide a snapshot of the outdoor air quality and the risks to human health that would result if air toxic emissions levels remained unchanged."

Most polluted cities by PM	
Particulate matter, μg/m³ (2004)	**City**
168	Cairo, Egypt
150	Delhi, India
128	Kolkata, India (Calcutta)
125	Tianjin, China
123	Chongqing, China
109	Kanpur, India
109	Lucknow, India
104	Jakarta, Indonesia
101	Shenyang, China

Governing Urban Air Pollution

In Europe, Council Directive 96/62/EC on ambient air quality assessment and management provides a common strategy against which member states can "set objectives for ambient air quality in order to avoid, prevent or reduce harmful effects on human health and the environment . . . and improve air quality where it is unsatisfactory".

On 25 July 2008 in the case Dieter Janecek v Freistaat Bayern CURIA, the European Court of Justice ruled that under this directive citizens have the right to require national authorities to implement a short term action plan that aims to maintain or achieve compliance to air quality limit values.

This important case law appears to confirm the role of the EC as centralised regulator to European nation-states as regards air pollution control. It places a supranational legal obligation on the UK to protect its citizens from dangerous levels of air pollution, furthermore superseding national interests with those of the citizen.

In 2010, the European Commission (EC) threatened the UK with legal action against the successive breaching of PM10 limit values. The UK government has identified that if fines are imposed, they could cost the nation upwards of £300 million per year.

In March 2011, the Greater London Built-up Area remains the only UK region in breach of the EC's limit values, and has been given 3 months to implement an emergency action plan aimed at meeting the EU Air Quality Directive. The City of London has dangerous levels of PM10 concentrations, estimated to cause 3000 deaths per year within the city. As well as the threat of EU fines, in 2010 it was threatened with legal action for scrapping the western congestion charge zone, which is claimed to have led to an increase in air pollution levels.

In response to these charges, Boris Johnson, Mayor of London, has criticised the current need for European cities to communicate with Europe through their nation state's central government, arguing that in future "A great city like London" should be permitted to bypass its government and deal directly with the European Commission regarding its air quality action plan.

This can be interpreted as recognition that cities can transcend the traditional national government organisational hierarchy and develop solutions to air pollution using global governance networks, for example through transnational relations. Transnational relations include but are not exclusive to national governments and intergovernmental organisations, allowing sub-national actors including cities and regions to partake in air pollution control as independent actors.

Particularly promising at present are global city partnerships. These can be built into networks, for example the C40 Cities Climate Leadership Group, of which London is a member. The C40 is a public 'non-state' network of the world's leading cities that aims to curb their greenhouse emissions. The C40 has been identified as 'governance from the middle' and is an alternative to intergovernmental policy. It has the potential to improve urban air quality as participating cities "exchange information, learn from best practices and consequently mitigate carbon dioxide emissions independently from national government decisions". A criticism of the C40 network is that its exclusive nature limits influence to participating cities and risks drawing resources away from less powerful city and regional actors.

Biogeochemical Cycle

In Earth science, a biogeochemical cycle or substance turnover or cycling of substances is a pathway by which a chemical substance moves through both the biotic (biosphere) and abiotic (lithosphere, atmosphere, and hydrosphere) components of Earth. A cycle is a series of change which comes back to the starting point and which can be repeated. Water, for example, is always recycled through the water cycle, as shown in the diagram. The water undergoes evaporation, condensation, and precipitation, falling back to Earth. Elements, chemical compounds, and other forms of matter are passed from one organism to another and from one part of the biosphere to another through biogeochemical cycles.'

The term "biogeochemical" tells us that biological, geological and chemical factors are all involved. The circulation of chemical nutrients like carbon, oxygen, nitrogen, phosphorus, calcium, and water etc. through the biological and physical world are known as biogeochemical cycles. In effect, the element is recycled, although in some cycles there may be places (called *reservoirs*) where the element is accumulated or held for a long period of time (such as an ocean or lake for water).

Important Cycles

The most well-known and important biogeochemical cycles, for example,

- the carbon cycle,
- the nitrogen cycle,
- the oxygen cycle,
- the phosphorus cycle,
- the sulfur cycle,
- the water cycle,
- and the rock cycle.

There are many biogeochemical cycles that are currently being studied for the first time as climate change and human impacts are drastically changing the speed, intensity, and balance of these relatively unknown cycles. These newly studied biogeochemical cycles include

- the mercury cycle, and
- the human-caused cycle of atrazine, which may affect certain species.

Biogeochemical cycles always involve hot equilibrium states: a balance in the cycling of the element between compartments. However, overall balance may involve compartments distributed on a global scale.

As biogeochemical cycles describe the movements of substances on the entire globe, the study of these is inherently multidisciplinary. The carbon cycle may be related to research in ecology and atmospheric sciences. Biochemical dynamics would also be related to the fields of geology and pedology (soil study).

Carbon Cycle

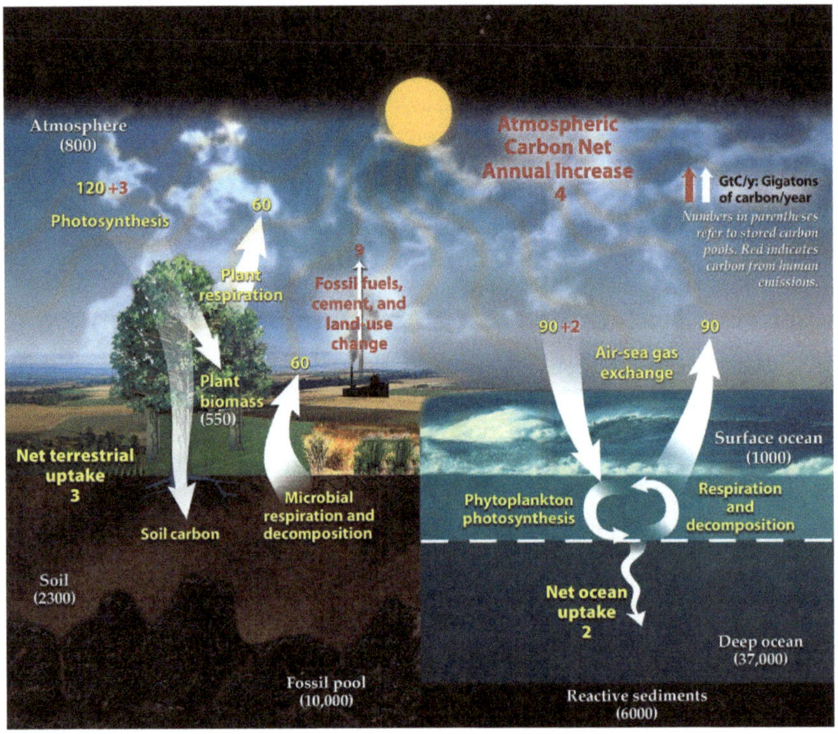

This diagram of the fast carbon cycle shows the movement of carbon between land, atmosphere, and oceans in billions of tons per year. Yellow numbers are natural fluxes, red are human contributions, white indicate stored carbon. This diagram does not account for volcanic and tectonic activity, which also sequesters and releases carbon.

The carbon cycle is the biogeochemical cycle by which carbon is exchanged among the biosphere, pedosphere, geosphere, hydrosphere, and atmosphere of the Earth. Along with the nitrogen cycle and the water cycle, the carbon cycle comprises a sequence of events that are key to make the Earth capable of sustaining life; it describes the movement of carbon as it is recycled and reused throughout the biosphere, including carbon sinks.

The *global carbon budget* is the balance of the exchanges (incomes and losses) of carbon between the carbon reservoirs or between one specific loop (e.g., atmosphere <-> biosphere) of the carbon cycle. An examination of the carbon budget of a pool or reservoir can provide information about whether the pool or reservoir is functioning as a source or sink for carbon dioxide. The carbon cycle was initially discovered by Joseph Priestley and Antoine Lavoisier, and popularized by Humphry Davy.

Global Climate

Carbon-based molecules are crucial for life on Earth, because it is the main component of biological compounds. Carbon is also a major component of many minerals. Carbon also exists in various forms in the atmosphere. Carbon dioxide (CO_2) is partly responsible for the greenhouse effect and is one of the most important human-contributed greenhouse gases.

In the past two centuries, human activities have altered the global carbon cycle, most significantly in the atmosphere. Although carbon dioxide levels have changed naturally over the past several thousand years, human emissions of carbon dioxide into the atmosphere have created unnatural fluctuations. Changes in the amount of atmospheric CO_2 are considerably altering weather patterns and indirectly influencing oceanic chemistry. Current carbon dioxide levels in the atmosphere exceed measurements from the last 420,000 years and levels are rising quickly, making it of critical importance to better understand how the carbon cycle works and what its effects are on the global climate.

Main Components

Carbon pools in the major reservoirs on earth.

Pool	Quantity (gigatons)
Atmosphere	720
Oceans (total)	38,400
Total inorganic	37,400
Total organic	1,000
Surface layer	670
Deep layer	36,730
Lithosphere	
Sedimentary carbonates	> 60,000,000
Kerogens	15,000,000
Terrestrial biosphere (total)	2,000
Living biomass	600 - 1,000
Dead biomass	1,200
Aquatic biosphere	1 - 2
Fossil fuels (total)	4,130
Coal	3,510
Oil	230
Gas	140
Other (peat)	250

The global carbon cycle is now usually divided into the following major reservoirs of carbon interconnected by pathways of exchange:

- The atmosphere
- The terrestrial biosphere
- The oceans, including dissolved inorganic carbon and living and non-living marine biota
- The sediments, including fossil fuels, fresh water systems and non-living organic material.
- The Earth's interior, carbon from the Earth's mantle and crust. These carbon stores interact with the other components through geological processes.

The carbon exchanges between reservoirs occur as the result of various chemical, physical, geological, and biological processes. The ocean contains the largest active pool of carbon near the surface of the Earth. The natural flows of carbon between the atmosphere, ocean, terrestrial ecosystems, and sediments is fairly balanced, so that carbon levels would be roughly stable without human influence.

Atmosphere

Epiphytes on electric wires. This kind of plant takes both CO_2 and water from the atmosphere for living and growing.

Carbon in the Earth's atmosphere exists in two main forms: carbon dioxide and methane. Both of these gases absorb and retain heat in the atmosphere and are partially responsible for the greenhouse effect. Methane produces a larger greenhouse effect per volume as compared to carbon dioxide, but it exists in much lower concentrations and is more short-lived than carbon dioxide, making carbon dioxide the more important greenhouse gas of the two.

Carbon dioxide leaves the atmosphere through photosynthesis, thus entering the terrestrial and oceanic biospheres. Carbon dioxide also dissolves directly from the atmosphere into bodies of water (oceans, lakes, etc.), as well as dissolving in precipitation as raindrops fall through the atmosphere. When dissolved in water, carbon dioxide reacts with water molecules and forms carbonic acid, which contributes to ocean acidity. It can then be absorbed by rocks through weathering. It also can acidify other surfaces it touches or be washed into the ocean.

Human activities over the past two centuries have significantly increased the amount of carbon in the atmosphere, mainly in the form of carbon dioxide, both by modifying ecosystems' ability to extract carbon dioxide from the atmosphere and by emitting it directly, e.g., by burning fossil fuels and manufacturing concrete.

Terrestrial Biosphere

A portable soil respiration system measuring soil CO_2 flux

The terrestrial biosphere includes the organic carbon in all land-living organisms, both alive and dead, as well as carbon stored in soils. About 500 gigatons of carbon are stored above ground in plants and other living organisms, while soil holds approximately 1,500 gigatons of carbon. Most carbon in the terrestrial biosphere is organic carbon, while about a third of soil carbon is stored in inorganic forms, such as calcium carbonate. Organic carbon is a major component of all organisms living on earth. Autotrophs extract it from the air in the form of carbon dioxide, converting it into organic carbon, while heterotrophs receive carbon by consuming other organisms.

Because carbon uptake in the terrestrial biosphere is dependent on biotic factors, it follows a diurnal and seasonal cycle. In CO_2 measurements, this feature is apparent in the Keeling curve. It is strongest in the northern hemisphere, because this hemisphere has more land mass than the southern hemisphere and thus more room for ecosystems to absorb and emit carbon.

Carbon leaves the terrestrial biosphere in several ways and on different time scales. The combustion or respiration of organic carbon releases it rapidly into the atmosphere. It can also be exported into the oceans through rivers or remain sequestered in soils in the form of inert carbon. Carbon stored in soil can remain there for up to thousands of years before being washed into rivers by erosion or released into the atmosphere through soil respiration. Between 1989 and 2008 soil respiration increased by about 0.1% per year. In 2008, the global total of CO_2 released from the soil reached roughly 98 billion tonnes, about 10 times more carbon than humans are now putting into the atmosphere each year by burning fossil fuel. There are a few plausible explanations for this trend, but the most likely explanation is that increasing temperatures have increased rates of decomposition of soil organic matter, which has increased the flow of CO_2. The length of carbon sequestering in soil is dependent on local climatic conditions and thus changes in the course of climate change. From pre-industrial era to 2010, the terrestrial biosphere represented a net source of atmospheric CO_2 prior to 1940, switching subsequently to a net sink.

Oceans

Oceans contain the greatest quantity of actively cycled carbon in this world and are second only to the lithosphere in the amount of carbon they store. The oceans' surface layer holds large amounts of dissolved inorganic carbon that is exchanged rapidly with the atmosphere. The deep layer's concentration of dissolved inorganic carbon (DIC) is about 15% higher than that of the surface layer. DIC is stored in the deep layer for much longer periods of time. Thermohaline circulation exchanges carbon between these two layers.

Carbon enters the ocean mainly through the dissolution of atmospheric carbon dioxide, which is converted into carbonate. It can also enter the oceans through rivers as dissolved organic carbon. It is converted by organisms into organic carbon through photosynthesis and can either be exchanged throughout the food chain or precipitated into the ocean's deeper, more carbon rich layers as dead soft tissue or in shells as calcium carbonate. It circulates in this layer for long periods of time before either being deposited as sediment or, eventually, returned to the surface waters through thermohaline circulation.

Oceanic absorption of CO_2 is one of the most important forms of carbon sequestering limiting the human-caused rise of carbon dioxide in the atmosphere. However, this process is limited by a number of factors. Because the rate of CO_2 dissolution in the ocean is dependent on the weathering of rocks and this process takes place slower than current rates of human greenhouse gas emissions, ocean CO_2 uptake will decrease in the future. CO_2 absorption also makes water more acidic, which affects ocean biosystems. The projected rate of increasing oceanic acidity could slow the biological precipitation of calcium carbonates, thus decreasing the ocean's capacity to absorb carbon dioxide.

Geological Carbon Cycle

The geologic component of the carbon cycle operates slowly in comparison to the other parts of the global carbon cycle. It is one of the most important determinants of the amount of carbon in the atmosphere, and thus of global temperatures.

Most of the earth's carbon is stored inertly in the earth's lithosphere. Much of the carbon stored in the earth's mantle was stored there when the earth formed. Some of it was deposited in the form of organic carbon from the biosphere. Of the carbon stored in the geosphere, about 80% is limestone and its derivatives, which form from the sedimentation of calcium carbonate stored in the shells of marine organisms. The remaining 20% is stored as kerogens formed through the sedimentation and burial of terrestrial organisms under high heat and pressure. Organic carbon stored in the geosphere can remain there for millions of years.

Carbon can leave the geosphere in several ways. Carbon dioxide is released during the metamorphosis of carbonate rocks when they are subducted into the earth's mantle. This carbon dioxide can be released into the atmosphere and ocean through volcanoes and hotspots. It can also be removed by humans through the direct extraction of kerogens in the form of fossil fuels. After extraction, fossil fuels are burned to release energy, thus emitting the carbon they store into the atmosphere.

Human Influence

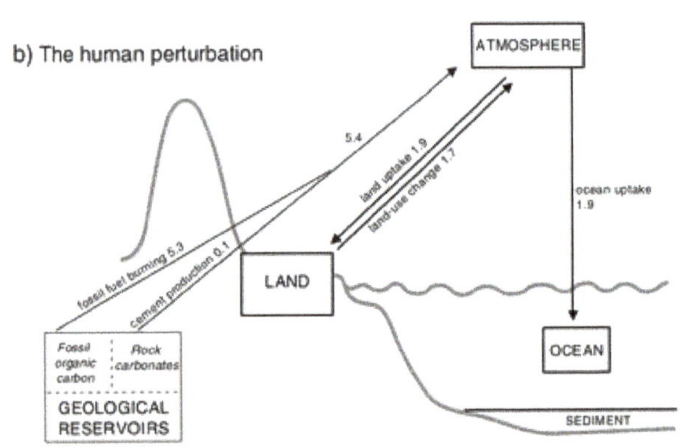

Human activity since the industrial era has changed the balance in the natural carbon cycle. Units are in gigatons.

Since the industrial revolution, human activity has modified the carbon cycle by changing its components' functions and directly adding carbon to the atmosphere. This interaction between humans and the carbon cycle can be described by carbon rift theory, which attributes the non-stop growth of carbon emissions to the capitalism.

The largest human impact on the carbon cycle is through direct emissions from burning fossil fuels, which transfers carbon from the geosphere into the atmosphere. The rest of this increase is caused mostly by changes in land-use, particularly deforestation.

Another direct human impact on the carbon cycle is the chemical process of calcination of limestone for clinker production, which releases CO_2. Clinker is an industrial precursor of cement.

CO_2 in Earth's atmosphere if *half* of global-warming emissions are *not* absorbed.(NASA computer simulation).

Humans also influence the carbon cycle indirectly by changing the terrestrial and oceanic biosphere. Over the past several centuries, direct and indirect human-caused land use and land cover change (LUCC) has led to the loss of biodiversity, which lowers ecosystems' resilience to environmental stresses and decreases their ability to remove carbon from the atmosphere. More directly, it often leads to the release of carbon from terrestrial ecosystems into the atmosphere. Deforestation for agricultural purposes removes forests, which hold large amounts of carbon, and replaces them, generally with agricultural or urban areas. Both of these replacement land cover types store comparatively small amounts of carbon, so that the net product of the process is that more carbon stays in the atmosphere.

Other human-caused changes to the environment change ecosystems' productivity and their ability to remove carbon from the atmosphere. Air pollution, for example, damages plants and soils, while many agricultural and land use practices lead to higher erosion rates, washing carbon out of soils and decreasing plant productivity.

Humans also affect the oceanic carbon cycle. Current trends in climate change lead to higher ocean temperatures, thus modifying ecosystems. Also, acid rain and polluted runoff from agriculture and industry change the ocean's chemical composition. Such changes can have dramatic effects on highly sensitive ecosystems such as coral reefs, thus limiting the ocean's ability to absorb carbon from the atmosphere on a regional scale and reducing oceanic biodiversity globally.

Arctic methane emissions indirectly caused by anthropogenic global warming also affect the carbon cycle, and contribute to further warming in what is known as climate change feedback.

On 12 November 2015, NASA scientists reported that human-made carbon dioxide (CO_2) continues to increase above levels not seen in hundreds of thousands of years: currently, about half of the carbon dioxide released from the burning of fossil fuels remains in the atmosphere and is not absorbed by vegetation and the oceans.

Nitrogen Cycle

The nitrogen cycle is the biogeochemical cycle by which nitrogen is converted into various chemical forms as it circulates among the atmosphere and terrestrial and marine ecosystems. The conversion of nitrogen can be carried out through both biological and physical processes. Important processes in the nitrogen cycle include fixation, ammonification, nitrification, and denitrification. The majority of Earth's atmosphere (78%) is nitrogen, making it the largest source of nitrogen. However, atmospheric nitrogen has limited availability for biological use, leading to a scarcity of usable nitrogen in many types of ecosystems. The nitrogen cycle is of particular interest to ecologists because nitrogen availability can affect the rate of key ecosystem processes, including primary production and decomposition. Human activities such as fossil fuel combustion, use of artificial nitrogen fertilizers, and release of nitrogen in wastewater have dramatically altered the global nitrogen cycle.

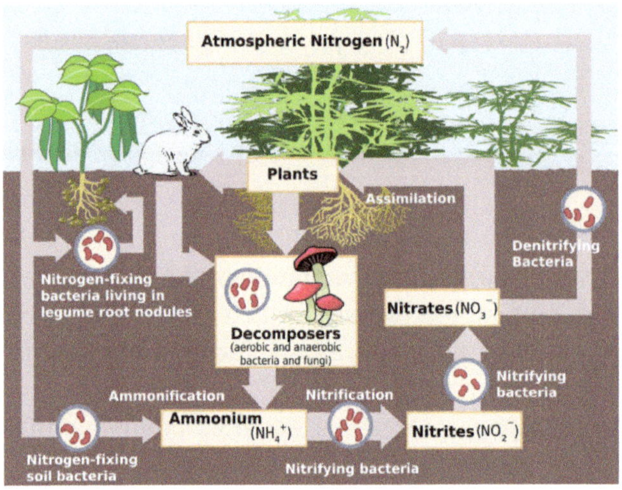

Schematic representation of the flow of nitrogen through the land environment. The importance of bacteria in the cycle is immediately recognized as being a key element in the cycle, providing different forms of nitrogen compounds assimilable by higher organisms

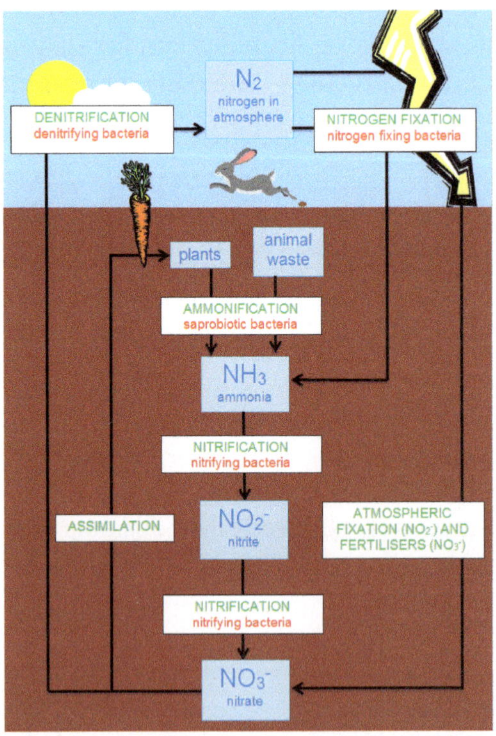

A simple diagram of the nitrogen cycle. The blue boxes represent stores of nitrogen, the green writing is for processes that occur to move the nitrogen from one place to another and the red writing are all the bacteria involved

The Processes of the Nitrogen Cycle

Nitrogen is present in the environment in a wide variety of chemical forms including organic nitrogen, Ammonium (NH^+_4), nitrite (NO^-_2), nitrate (NO^-_3), nitrous oxide (N_2O), Nitric oxide (NO) or inorganic nitrogen gas (N_2). Organic nitrogen may be in the form of a living organism, humus or in the intermediate products of organic matter decomposition. The processes of the nitrogen cycle transform nitrogen from one form to another. Many of those processes are carried out by microbes, either in their effort to harvest energy or to accumulate nitrogen in a form needed for their growth. For example, the nitrogenous wastes in animal urine are broken down by nitrifying bacteria in the soil to be used anew. The diagram above shows how these processes fit together to form the nitrogen cycle.

Nitrogen Fixation

Atmospheric nitrogen must be processed, or "fixed", in a usable form to be taken up by plants. Between $5x10^{12}$ and $10x10^{12}$ g per year are fixed by lightning strikes, but most fixation is done by free-living or symbiotic bacteria known as diazotrophs. These bacteria have the nitrogenase enzyme that combines gaseous nitrogen with hydrogen to produce ammonia, which is converted by the bacteria into other organic compounds. Most biological nitrogen fixation occurs by the activity of Mo-nitrogenase, found in a wide variety of bacteria and some Archaea. Mo-nitrogenase is a complex two-component enzyme that has multiple metal-containing prosthetic groups. An example of the free-living bacteria is *Azotobacter*. Symbiotic nitrogen-fixing bacteria such as *Rhizobium* usually live in the root nodules of legumes (such as peas, alfalfa, and locust trees). Here they form a mutualistic relationship with the plant, producing ammonia in exchange for carbohydrates. Because of this relationship, legumes will often increase the nitrogen content of nitrogen-poor soils. A few non-legumes can also form such symbioses. Today, about 30% of the total fixed nitrogen is produced industrially using the Haber-Bosch process, which uses high temperatures and pressures to convert nitrogen gas and a hydrogen source (natural gas or petroleum) into ammonia.

Assimilation

Plants take nitrogen from the soil by absorption through their roots as amino acids, nitrate ions, nitrite ions, or ammonium ions. Most nitrogen obtained by terrestrial animals can be traced back to the eating of plants at some stage of the food chain.

Plants can absorb nitrate or ammonium from the soil via their root hairs. If nitrate is absorbed, it is first reduced to nitrite ions and then ammonium ions for incorporation into amino acids, nucleic acids, and chlorophyll. In plants that have a symbiotic relationship with rhizobia, some nitrogen is assimilated in the form of ammonium ions directly from the nodules. It is now known that there is a more complex cycling of amino acids between *Rhizobia* bacteroids and plants. The plant provides amino acids to the bacteroids so ammonia assimilation is not required and the bacteroids pass amino acids (with the newly fixed nitrogen) back to the plant, thus forming an interdependent relationship. While many animals, fungi, and other heterotrophic organisms obtain nitrogen by ingestion of amino acids, nucleotides, and other small organic molecules, other heterotrophs (including many bacteria) are able to utilize inorganic compounds, such as ammonium as sole N sources. Utilization of various N sources is carefully regulated in all organisms.

Ammonification

When a plant or animal dies or an animal expels waste, the initial form of nitrogen is organic. Bacteria or fungi convert the organic nitrogen within the remains back into ammonium (NH+4), a process called ammonification or mineralization. Enzymes involved are:

- GS: Gln Synthetase (Cytosolic & Plastic)
- GOGAT: Glu 2-oxoglutarate aminotransferase (Ferredoxin & NADH-dependent)
- GDH: Glu Dehydrogenase:
    - Minor Role in ammonium assimilation.
    - Important in amino acid catabolism.

Nitrification

The conversion of ammonium to nitrate is performed primarily by soil-living bacteria and other nitrifying bacteria. In the primary stage of nitrification, the oxidation of ammonium (NH^+_4) is performed by bacteria such as the *Nitrosomonas* species, which converts ammonia to nitrites (NO^-_2). Other bacterial species such as *Nitrobacter*, are responsible for the oxidation of the nitrites (NO^-_2) into nitrates (NO^-_3). It is important for the ammonia (NH_3) to be converted to nitrates or nitrites because ammonia gas is toxic to plants.

Due to their very high solubility and because soils are highly unable to retain anions, nitrates can enter groundwater. Elevated nitrate in groundwater is a concern for drinking water use because nitrate can interfere with blood-oxygen levels in infants and cause methemoglobinemia or blue-baby syndrome. Where groundwater recharges stream flow, nitrate-enriched groundwater can contribute to eutrophication, a process that leads to high algal population and growth, especially blue-green algal populations. While not directly toxic to fish life, like ammonia, nitrate can have indirect effects on fish if it contributes to this eutrophication. Nitrogen has contributed to severe eutrophication problems in some water bodies. Since 2006, the application of nitrogen fertilizer has been increasingly controlled in Britain and the United States. This is occurring along the same lines as control of phosphorus fertilizer, restriction of which is normally considered essential to the recovery of eutrophied waterbodies.

Denitrification

Denitrification is the reduction of nitrates back into nitrogen gas (N_2), completing the nitrogen cycle. This process is performed by bacterial species such as *Pseudomonas* and *Clostridium* in anaerobic conditions. They use the nitrate as an electron acceptor in the place of oxygen during respiration. These facultatively anaerobic bacteria can also live in aerobic conditions. Denitrification happens in anaerobic conditions e.g. waterlogged soils. The denitrifying bacteria use nitrates in the soil to carry out respiration and consequently produce nitrogen gas, which is inert and unavailable to plants.

Anaerobic Ammonia Oxidation

In this biological process, nitrite and ammonia are converted directly into molecular nitrogen (N_2) gas. This process makes up a major proportion of nitrogen conversion in the oceans. The balanced formula for this "anammox" chemical reaction is: $NH_4^+ + NO_2^- => N_2 + 2H_2O$ ($\Delta G° = -357$kJ mol-1).

Other Processes

Though nitrogen fixation is the primary source of plant-available nitrogen in most ecosystems, in areas with nitrogen-rich bedrock, the breakdown of this rock also serves as a nitrogen source.

Marine Nitrogen Cycle

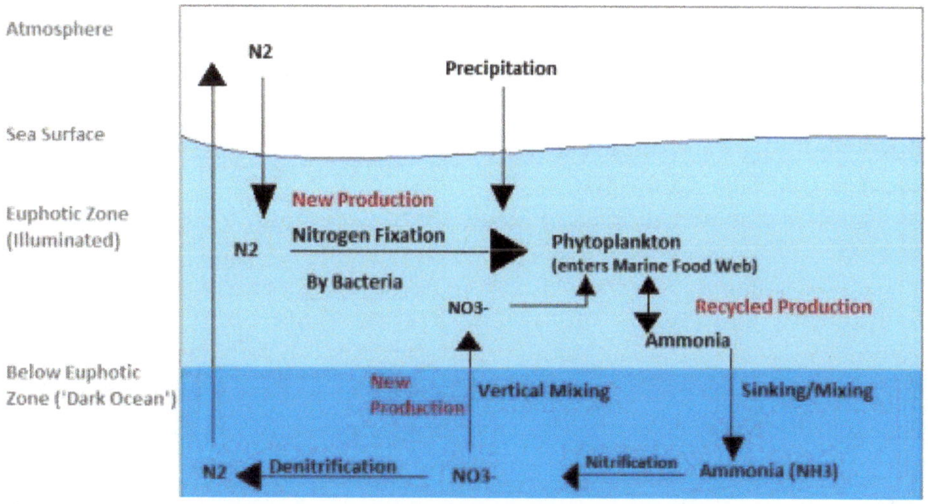

A schematic representing the Marine Nitrogen Cycle

The nitrogen cycle is an important process in the ocean as well. While the overall cycle is similar, there are different players and modes of transfer for nitrogen in the ocean. Nitrogen enters the water through precipitation, runoff, or as N_2 from the atmosphere. Nitrogen cannot be utilized by phytoplankton as N_2 so it must undergo nitrogen fixation which is performed predominately by cyanobacteria. Without supplies of fixed nitrogen entering the marine cycle, the fixed nitrogen would be used up in about 2000 years. Phytoplankton need nitrogen in biologically available forms for the initial synthesis of organic matter. Ammonia and urea are released into the water by excretion from plankton. Nitrogen sources are removed from the euphotic zone by the downward movement of the organic matter. This can occur from sinking of phytoplankton, vertical mixing, or sinking of waste of vertical migrators. The sinking results in ammonia being introduced at lower depths below the euphotic zone. Bacteria are able to convert ammonia to nitrite and nitrate but they are inhibited by light so this must occur below the euphotic zone. Ammonification or Mineralization is performed by bacteria to convert organic nitrogen to ammonia. Nitrification can then occur to convert the ammonium to nitrite and nitrate. Nitrate can be returned to the euphotic zone by vertical mixing and upwelling where it can be taken up by phytoplankton to continue the cycle. N_2 can be returned to the atmosphere through denitrification.

Ammonium is thought to be the preferred source of fixed nitrogen for phytoplankton because its assimilation does not involve a redox reaction and therefore requires little energy. Nitrate requires a redox reaction for assimilation but is more abundant so most phytoplankton have adapted to have the enzymes necessary to undertake this reduction (nitrate reductase). There are a few notable and well-known exceptions that include Prochlorococcus and some Synechococcus. These species can only take up nitrogen as ammonium.

The nutrients in the ocean are not uniformly distributed. Areas of upwelling provide supplies of nitrogen from below the euphotic zone. Coastal zones provide nitrogen from runoff and upwelling occurs readily along the coast. However, the rate at which nitrogen can be taken up by phytoplankton is decreased in oligotrophic waters year-round and temperate water in the summer resulting in lower primary production. The distribution of the different forms of nitrogen varies throughout the oceans as well.

Nitrate is depleted in near-surface water except in upwelling regions. Coastal upwelling regions usually have high nitrate and chlorophyll levels as a result of the increased production. However, there are regions of high surface nitrate but low chlorophyll that are referred to as HNLC (high nitrogen, low chlorophyll) regions. The best explanation for HNLC regions relates to iron scarcity in the ocean, which may play an important part in ocean dynamics and nutrient cycles. The input of iron varies by region and is delivered to the ocean by dust (from dust storms) and leached out of rocks. Iron is under consideration as the true limiting element to ecosystem productivity in the ocean.

Ammonium and nitrite show a maximum concentration at 50–80 m (lower end of the euphotic zone) with decreasing concentration below that depth. This distribution can be accounted for by the fact that nitrite and ammonium are intermediate species. They are both rapidly produced and consumed through the water column. The amount of ammonium in the ocean is about 3 orders of magnitude less than nitrate. Between ammonium, nitrite, and nitrate, nitrite has the fastest turnover rate. It can be produced during nitrate assimilation, nitrification, and denitrification; however, it is immediately consumed again.

New vs. Regenerated Nitrogen

Nitrogen entering the euphotic zone is referred to as new nitrogen because it is newly arrived from outside the productive layer. The new nitrogen can come from below the euphotic zone or from outside sources. Outside sources are upwelling from deep water and nitrogen fixation. If the organic matter is eaten, respired, delivered to the water as ammonia, and re-incorporated into organic matter by phytoplankton it is considered recycled/regenerated production. New production is an important component of the marine environment. One reason is that only continual input of new nitrogen can determine the total capacity of the ocean to produce a sustainable fish harvest. Harvesting fish from regenerated nitrogen areas will lead to a decrease in nitrogen and therefore a decrease in primary production. This will have a negative effect on the system. However, if fish are harvested from areas of new nitrogen the nitrogen will be replenished.

Human Influences on the Nitrogen Cycle

As a result of extensive cultivation of legumes (particularly soy, alfalfa, and clover), growing use of the Haber–Bosch process in the creation of chemical fertilizers, and pollution emitted by vehicles and industrial plants, human beings have more than doubled the annual transfer of nitrogen into biologically available forms. In addition, humans have significantly contributed to the transfer of nitrogen trace gases from Earth to the atmosphere and from the land to aquatic systems. Human alterations to the global nitrogen cycle are most intense in developed countries and in Asia, where vehicle emissions and industrial agriculture are highest.

Nitrous oxide (N_2O) has risen in the atmosphere as a result of agricultural fertilization, biomass burning, cattle and feedlots, and industrial sources. N_2O has deleterious effects in the stratosphere, where it breaks down and acts as a catalyst in the destruction of atmospheric ozone. Nitrous oxide is also a greenhouse gas and is currently the third largest contributor to global warming, after carbon dioxide and methane. While not as abundant in the atmosphere as carbon dioxide, it is, for an equivalent mass, nearly 300 times more potent in its ability to warm the planet.

Ammonia (NH_3) in the atmosphere has tripled as the result of human activities. It is a reactant in the atmosphere, where it acts as an aerosol, decreasing air quality and clinging to water droplets, eventually resulting in nitric acid (HNO_3) that produces acid rain. Atmospheric ammonia and nitric acid also damage respiratory systems.

The very-high temperature of lightning naturally produces small amounts of NO_x, NH_3, and HNO_3, but high-temperature combustion has contributed to a 6 or 7 fold increase in the flux of NO_x to the atmosphere. Its production is a function of combustion temperature - the higher the temperature, the more NO_x is produced. Fossil fuel combustion is a primary contributor, but so are biofuels and even the burning of hydrogen. The higher combustion temperature of hydrogen produces more NO_x than natural gas combustion.

Ammonia and nitrous oxides actively alter atmospheric chemistry. They are precursors of tropospheric (lower atmosphere) ozone production, which contributes to smog and acid rain, damages plants and increases nitrogen inputs to ecosystems. Ecosystem processes can increase with nitrogen fertilization, but anthropogenic input can also result in nitrogen saturation, which weakens productivity and can damage the health of plants, animals, fish, and humans.

Decreases in biodiversity can also result if higher nitrogen availability increases nitrogen-demanding grasses, causing a degradation of nitrogen-poor, species diverse heathlands.

Environmental Impacts

Additional risks posed by increased availability of inorganic nitrogen in aquatic ecosystems include water acidification; eutrophication of fresh and saltwater systems; and toxicity issues for animals, including humans. Eutrophication often leads to lower dissolved oxygen levels in the water column, including hypoxic and anoxic conditions, which can cause death of aquatic fauna. Relatively sessile benthos, or bottom-dwelling creatures, are particularly vulnerable because of their lack of mobility, though large fish kills are not uncommon. Oceanic dead zones near the mouth of the Mississippi in the Gulf of Mexico are a well-known example of algal bloom-induced hypoxia. The New York Adirondack Lakes, Catskills, Hudson Highlands, Rensselaer Plateau and parts of Long Island display the impact of nitric acid rain deposition, resulting in the killing of fish and many other aquatic species.

Ammonia (NH_3) is highly toxic to fish and the level of ammonia discharged from wastewater treatment facilities must be closely monitored. To prevent fish deaths, nitrification via aeration prior to discharge is often desirable. Land application can be an attractive alternative to the aeration.

Sulfur Cycle

Sulfur reservoirs

Driving forces for sulfur transformation

Sulfur in the atmosphere

Sulfur in rivers

Sulfur in the ocean

Global sulfur cycle

Sulfur budget for the ocean

Pyrite formation in sediments

Diagenetic modelling

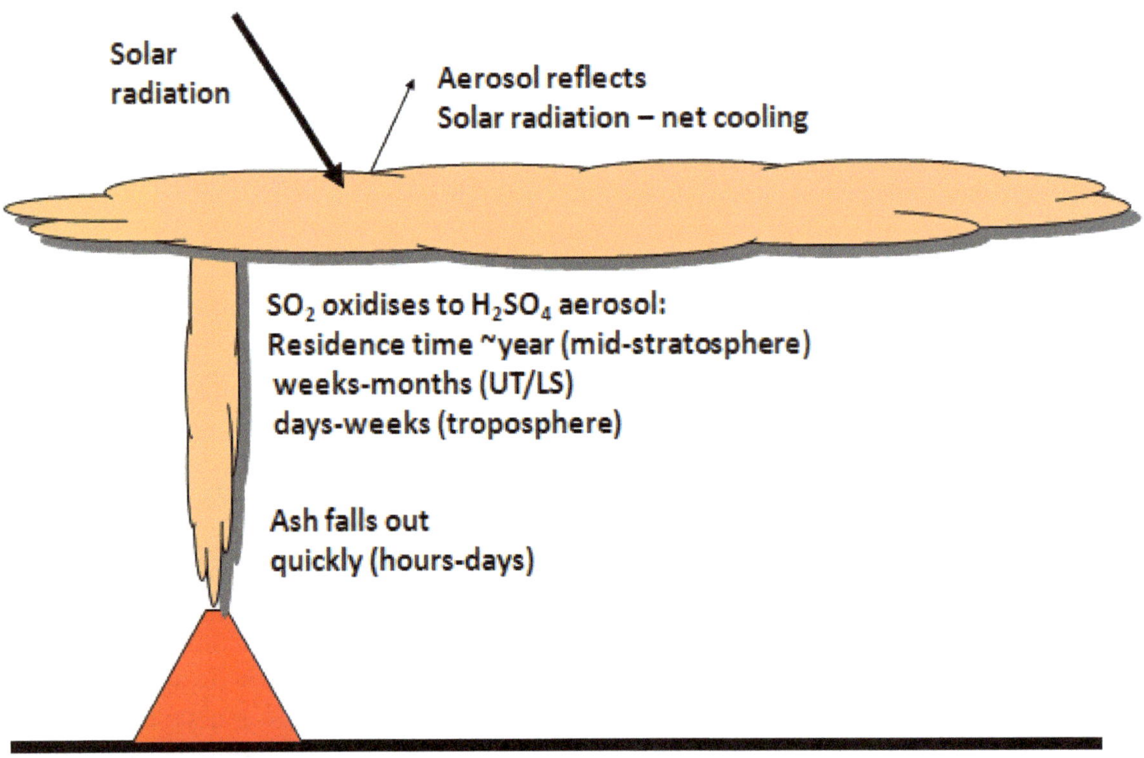

Sulphur Cycle

$$SO_{2\,(gas)} + OH \rightarrow H_2SO_{4\,(aerosol)}$$

Global Sulfur Cycle

- Valence states: +6 (SO_4^{2-}) to -2 (sulfides)
- Original pool - pyrite FeS_2
- Reservoirs,1018 g S:

Deep oceanic rocks 2375 ±820

Sediments 7800±1700

Freshwater 0.003±0.002

Ice 0.006±0.002

Atmosphere 3.6

Sea 1280±55

Organic 5.62x 10^{-3}

Driving Forces: Microbial Transformation

Anaerobic conditions:

sulfate reduction: $2H^+ + SO_4^{2-} + 2(CH_2O) -> 2CO_2 + H_2S + 2H_2O$ (Desulfovibrio sp. or Desulfotomaculum sp.))

- bacteria produce a variety of gases: hydrogen sulfide (H_2S), dimethyl-sulfide $(CH_3)_2S$, carboxyl sulfide COS
- H_2S reacts with Fe^{2+} to precipitate FeS, which can be converted to pyrite FeS_2:

$$FeS + H_2S > FeS_2 + 2H^+ + 2e^-$$

- H_2S diffuses though zone of F^{3+}

$$2Fe(OH)_3 + 3H_2S + 2H^+ > FeS_2 + 6H_2O + Fe^{2+}$$

sulfur-based photosynthesis (thought to be one of first forms of photosynthesis on the Earth)

Plant uptake:

assimilatory SO_4^{2-} reduction and incorporation of carbon-bounded sulfur into the amino acids cysteine and methionine.

Aerobic conditions:

reduced sulfur compounds oxidized by microbes, oxidation usually coupled to reduction of CO_2 in relations of S-based chemosynthesis.

Sulfur in the Atmosphere

- Gaseous component
  - no sulfur gas is a long-lived or major constituent of the atmosphere, oxidation of SO_4^{2-} → short residence time, all expressions in g S
- Aerosols
  - particles < 1um are held a loft by Brownian motion
  - long transport
  - sources: volcanic eruptions, ocean, water evaporates from bubbles, the salt crystallizes to form sea-salt aerosols

Sources of Sulfur in the Atmosphere

Eriksson (1960) - SO_4^{2-} deposition on land from ocean $4x10^{12}$ g S ,

Jung (1960) - SO_4^{2-} in rainfall in land $73x10^{12}$ g S, → other sources as sea.

- Volcanic eruptions
  - average over many years $12\text{-}30x10^{12}$ g S
  - e.g. Tambora (Indonesia) in 1815, 1816 - year without summer in England, USA, Canada, $50x10^{12}$ g S
- Soil dust
- Biogenic gases
  - H_2S, dimethyl-sulfide $(CH_3)_2S$, carbonyl sulfide COS
- Anthropogenic emissions
  - without human effects, net transport through the atmosphere carries S from sea to land

Sulfur in Rivers

- Natural river load
  - o from weathering of pyrite

$$4FeS_2 + 15O_2 + 8H2O \rightarrow 2Fe_2O_3 + 8H_2SO_4$$

  - o and gypsum, rainfall
- Human activities affect the transport of S in rivers
  - o 28% of the current content of S in rivers is derived from air pollution, mining, erosion, and the other human activities
  - o the current transport is supposed to be about double that of pre-industrial conditions

Marine Sulfur Cycle

- Ocean is large source of aerosols (sea salts) that contains SO_4^{2-}.
- Most of the flux is re-deposited in the ocean in precipitation and dry-fall
- Dimethyl-sulfide $(CH_3)_2S$ or DMS is the major biogenic gases emitted from sea
- annual flux is about 15
- mean residence time about 1-2 days - most of S from DMS is also re-deposited in the ocean
- The net transport of S from sea to land is about 20x10[12] g S/yr. Ocean receives a net input of S.

Global Sulfur Cycle

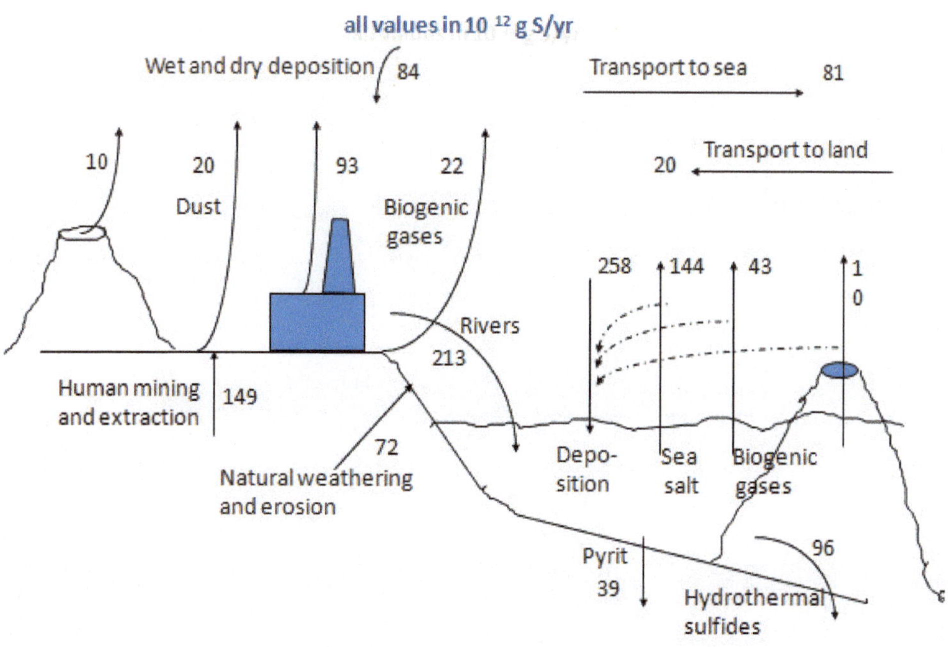

Sulfur Budget for the Ocean

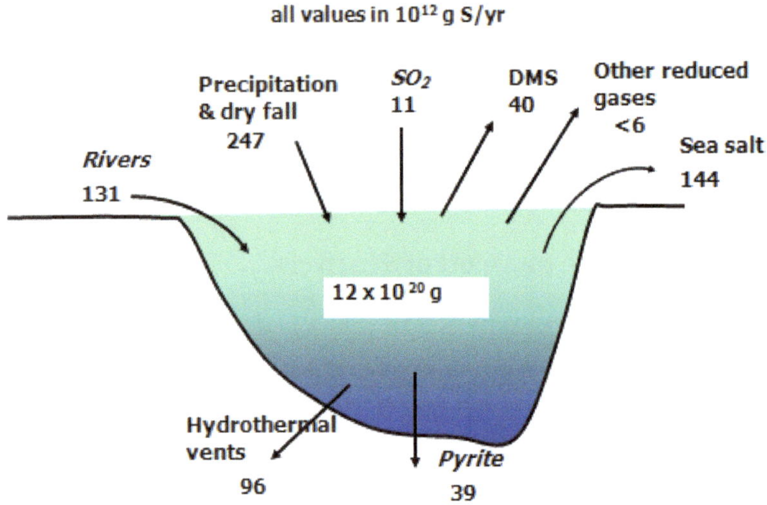

Marine Sulfur Cycle

- Content $12x10^{20}$ g S/ yr., residence time > 3 000 000 years.
- Major marine sinks: metallic sulfides precipitated at hydro-thermal vents, biogenic pyrites, the formation of organic sulfur.

Dimethylsulfid$(CH_3)_2S$ or DMS

DMS is the major one of biogenic gases emitted from sea

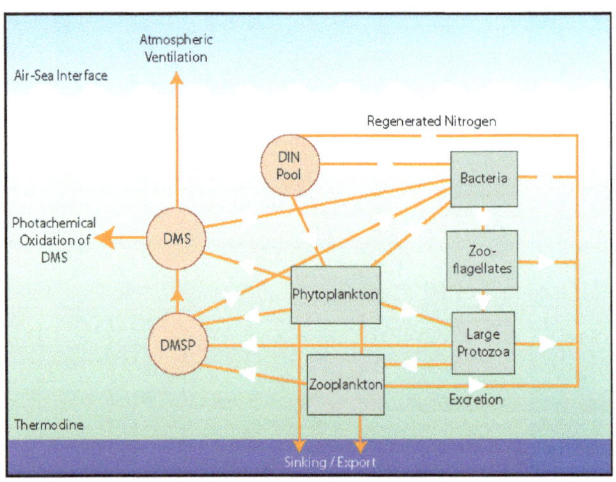

- Mean residence time is about 1-2 days - most of S from DMS is also re-deposited in the ocean
- Is produced during decom-position of dimethyl-sulfonpropionate (DMSP) from dying phytoplankton
- Only small fraction lost into the atmosphere

DMS and Climate

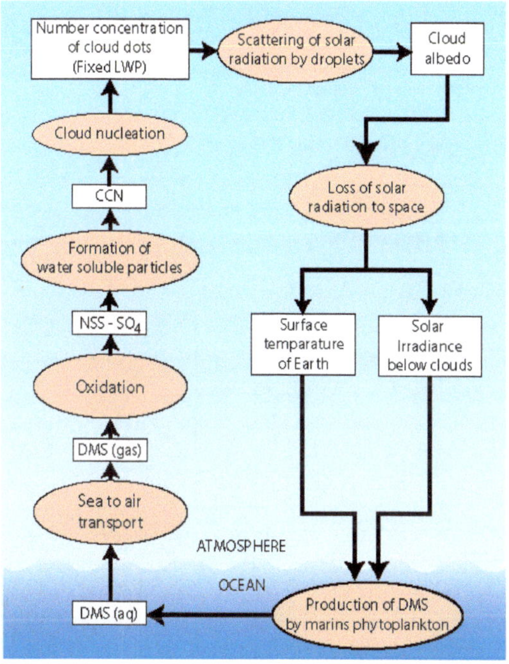

- oxidation of DMS to sulfate aerosols increases the abundance of cloud condensation nuclei → to greater cloudiness
- layer of sulfate aerosols (known as Junge layer) is about 20-25 km altitude, source: SO_2 and carbonyl sulfate COS
- clouds over sea reflect incoming sunlight → global cooling
- production of DMS - net primary production.
- if higher NPP is associated with warmer sea surface, then DMS-flux would have negative feedback on global warming

Pyrite Formation in Sediments

Sulfate reduction : $SO_4^{2-} + 2(CH_2O) -> 2HCO_3^- + H_2S$

Two Steps Reaction

- Reaction H_2S with Fe^{2+} or reactive Fe-mineral

$4Fe_2O_3 + 9\mathbf{H_2S} -> 8FeS + SO_4^{2-} + 8H_2O + 2H^+$

- Reaction of iron sulfide with elemental sulfur

$FeS + S^0 -> FeS_2$

incomplete oxidation of H_2S or FeS by O_2, NO_3^-, MnO_2 or FeOOH

Recently

In strictly anoxic sediments $FeS + H2S -> FeS_2 + H_2$

In salt sediments $Fe^{2+} + S^0 -> FeS_2$

Pyrite Formation in Sediments (After Berner 1972)

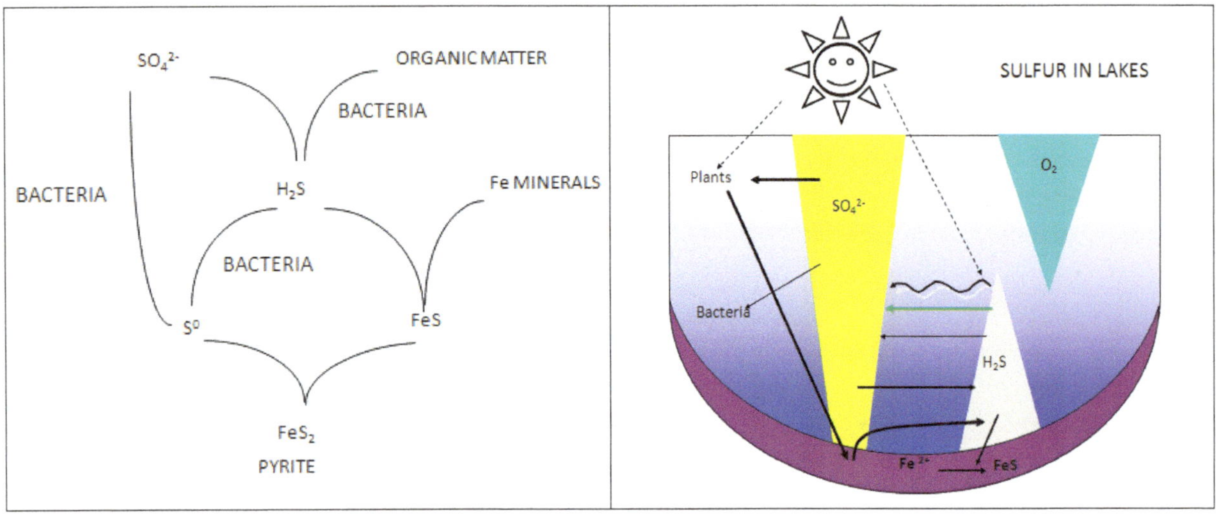

Organic Pollutants

Persistent Organic Pollutants

Humans are exposed daily to numerous chemicals that can harm their health

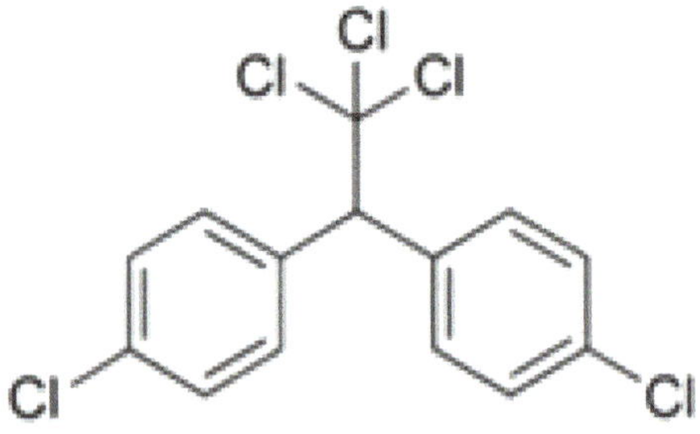

Many harmful organic compounds are stable in the environment (atmosphere, water, soil, food chain) for long periods → Persistent Organic Pollutants (POPs)

- Characterised by stability, mobility, and bioaccumulation
- Harmful to human health and produce ecological damage

Stockholm Convention (May 2001): over 90 countries promised to reduce or eliminate the production, use, and release of 12 key POPs – the “dirty dozen”.

12 Key POPs – The Dirty Dozen

The Dirty Dozen are produced for use as insecticides, fungicides, chemical additives, or are inadvertently produced during combustion. A biocide is a substance toxic to varying degrees to life forms:

- Either synthesised deliberately to target and kill specific organisms – general name pesticide
- Or may be inadvertently produced

Pesticides: insecticide, bactericide, fungicide, herbicide

A pesticide designed to eliminate all types of living organisms is called a fumigant or sterilant

POP	Use	Structure
Aldrin	Crop insecticide (corn, cotton)	
Chlordane	Crop insecticide (vegetables, citrus, cotton, potatoes)	
DDT (dichloro-diphenyl-trichloroethane)	Crop insecticide (cotton)	
Dieldrin	Crop insecticide (cotton, corn)	
Endrin	Crop insecticide (cotton, grains)	
Heptachlor	Insecticide (termites and soil insects)	
Hexachlorobenzene	Fungicide for seed treatment	
Mirex	Insecticide (termites, fire ants)	
Toxaphene	Insecticide (livestock, crops)	
PCBs	Industrial chemical (paint and plastic additive)	
Dioxins	Unintentionally produced during combustion	
Furans	Unintentionally produced during combustion	

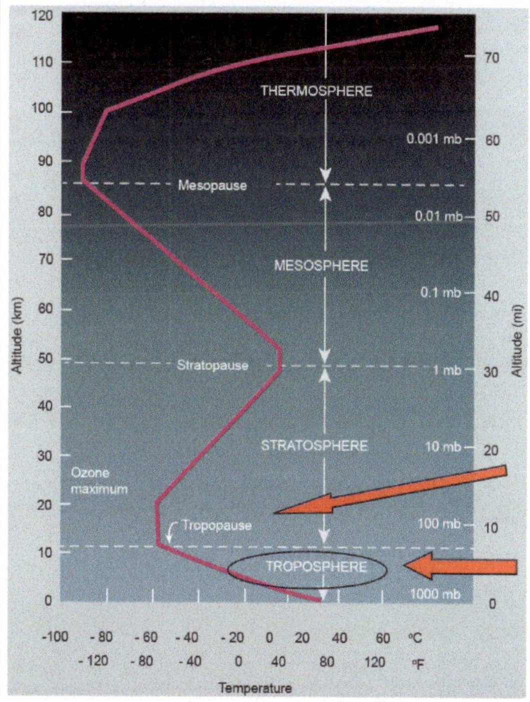

Organic Biocides

Fate of biocides in the environment:

- Chemical stability (rate it degrades to other compounds)
- Mobility (mechanism and rate of transport)

Rapid degradation mobility less of an issue

Different degradation mechanisms may operate as the pesticide moves to a new environment

Degradation products may have biocidal properties:

- Usually reduced toxicity, but sometimes toxicity is enhanced

Polychlorinated Hydrocarbons

All the Dirty Dozen are polychlorinated hydrocarbons

They are stable and do not readily degrade to less toxic forms

- Persist in the environment for a long time and can bioaccumulate through the food chains

Insoluble in water, soluble in fats

Polychlorinated hydrocarbons of particular concern are:

DDT

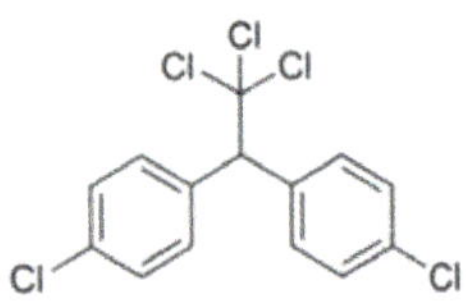

dioxins

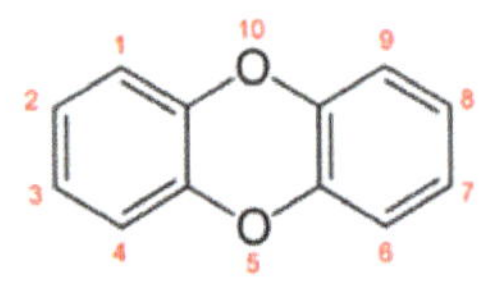

PCBs

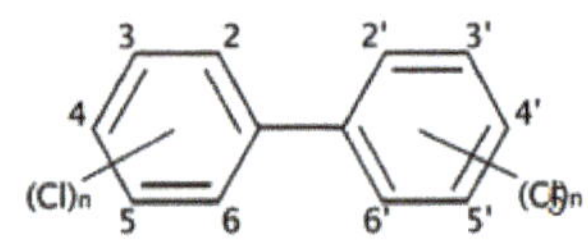

Pesticide Production

Few pesticides were available before 1940:

- Some natural insect poisons extracted from plants
- Early insecticides were nicotine sulfate obtained from tobacco, retenone from the tropical derris plant and garlic oil
- A few inorganic compounds of lead and arsenic

Large scale use of pesticides began after WWII with the introduction of DDT

DDT

At its introduction it seemed to be an ideal insecticide:

- Cheap to produce
- Apparently non-toxic to humans and other mammals
- Did not breakdown easily → continued to kill insects for a long time after application

DDT – A Short History

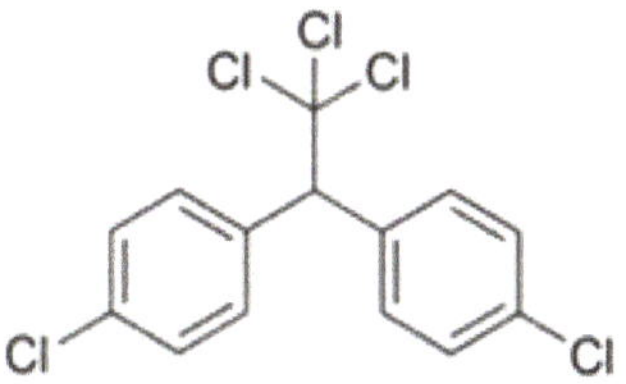

1874 – First synthesised by Othmar Zeidler - dichlorodiphenyltrichloroethane (DDT)

1939 – Paul Hermann Müller discovered insecticidal properties of DDT

1940s – Rapid development in WW2 to combat malaria, typhus, and other mosquito-borne diseases in military & civilian populations

- Following WW2, extensive use of DDT contributed to eradication of malaria in America & Europe

1948 – Müller awarded Nobel Prize

1950s – A WHO campaign to eradicate malaria was based on extensive spraying with DDT – a marked reduction in malaria mortality rate was found

→ mortality rate dropped from 192 to 7 per 100,000

- increasing resistance to DDT was found in pests

1950s – Marked decline of bird populations, especially raptors like the peregrine falcon, osprey, and eagles, were observed

1962 – Rachel Carson publishes Silent Spring, raising American awareness of the ecological effects & health risks of DDT and start of modern environmental movement.

1970s-80s – ban on DDT use in Western countries; recovery of affected bird populations

- DDT use continued in developing countries, but widespread spraying discontinued. Insecticide impregnated bednets and indoor spraying favoured.

2001 – Stockholm convention, restricting DDT use to health crisis.

DDT

Chemical stability: DDT degrades to DDE (dichlorodiphenyldichloroethene)

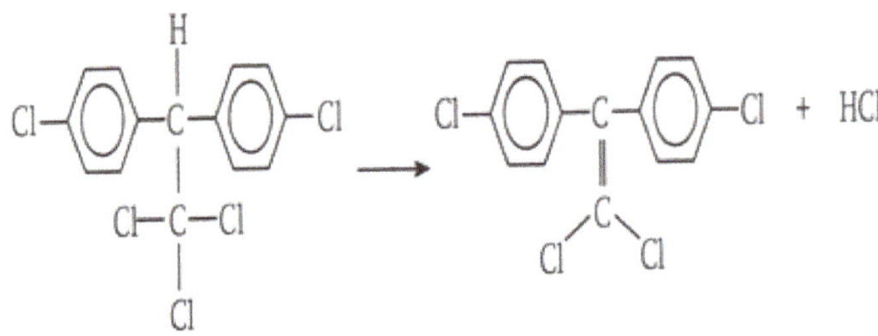

DDT DDE

DDE is less toxic than DDT but more resilient in the environment

DDT has a broad spectrum of activity. It was banned in several countries in the 1970s because of ecological considerations

Still used extensively to reduce insect-transmitted diseases:

- yellow fever, sleeping sickness, typhus, malaria and others

Persistent in the environment and resistant to complete degradation by microorganisms Photo-degradation can occur

Readily adsorbed onto sediments and soil

Most DDT that enters water becomes firmly attached to soil particles

Readily taken up by microorganisms and is therefore passed on through the food chain

Normally stored in fatty tissue

The Effect of DDT on Ecosystems

E.g., DDT in Lake Kariba
(Zimbabwe, Zambia)

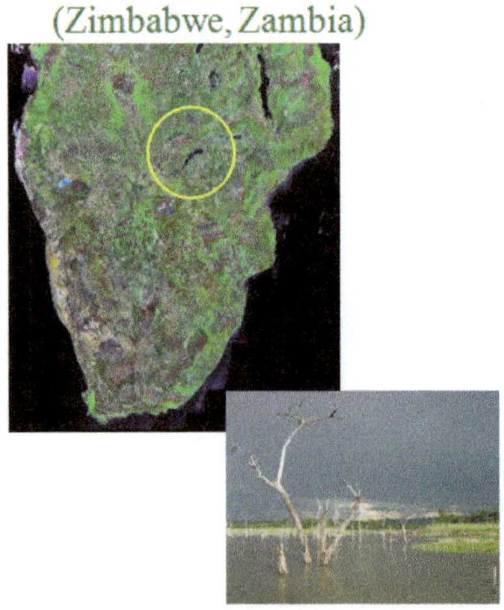

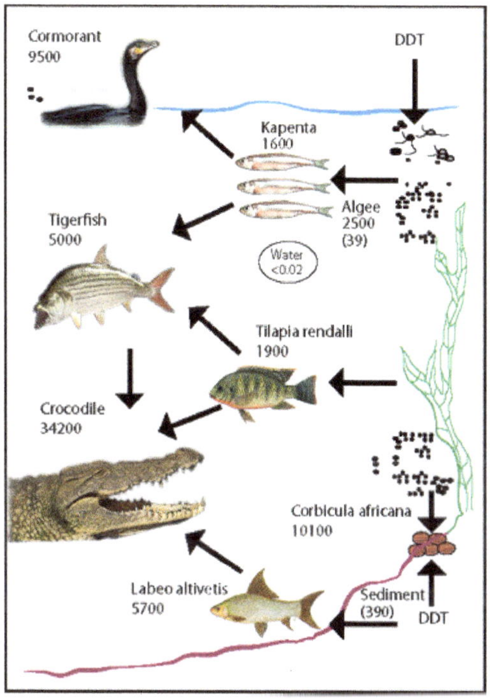

Mean levels of DDT (ng/g in fat) in the Lake Kabira ecosystem (Berg. 1992).

DDT – Environmental Levels and Human Exposure

Air: when DDT is sprayed, any that fails to reach its target can drift away. Vaporisation from treated fields can be detected for more than 6 months after application. It can drift up to 1000 km

Environmental levels:

- Nonagricultural areas: $< 1 - 2.4$ ng m^{-3}
- Agricultural communities: $1 - 22$ ng m^{-3}
- Communities with anti-mosquito programmes: up to 8.5 μg m^{-3}

Water: highest level recorded 0.84 mg/L (USA: 1964 – 1968)

- Over 90% of DDT in the general population comes from food

Acute exposure: Acute toxicity of DDT is high in insects and lower in mammals. Large doses causes focal necrosis of liver cells in several species. It increases liver tumours in mice

Human exposure: Can affect the nervous system. Acute intoxication by DDT can lead to symptoms such as nausea, vomiting, paraesthsia, dizziness, confusion, tremors and in severe cases convulsions

All the symptoms are rare

No evidence that DDT has reproductive or teratogenic effects

All epidemiological studies in humans have indicated that DDT is not carcinogenic

Combustion Released Pollutants

Pollutant Formation in Combustion Systems

- NO_x Formation

  Thermal NO_x

  Prompt NO_x

  Fuel NO_x

- Carbon Monoxide (CO)
- Volatile Organic Compounds (VOC)
- Polycyclic Aromatic Hydrocarbons (PAH), Soot and Submicron Particulates (solid-phase pollutants)
- Sulphur Compounds

NOx Formation

What is NO_x?

- NO_x = Oxides of Nitrogen which are produced by combustion:

  Nitric Oxide (NO)

  Nitrogen Dioxide (NO_2)

  Nitrous Oxide (N_2O)

Nitric Oxide (NO)

- Highly reactive due to lone electron at N atom
- Not particularly toxic
- Major precursor of photochemical smog ($NO \rightarrow NO_2$)
- It is produced by most of combustion systems

Nitrogen Dioxide (NO_2)

- Brown, poisonous gas
- Emissions of NO_2 from most combustion < 10% of NO_x
- Adverse health effects include: lung irritation, bronchitis, pneumonia and a lowering respiratory resistance
- Ambient limit = 120ppb [NEPC, 1998]
- Significant direct emissions of NO_2 occur from processes involving premixed flames:

- Indoor gas appliances (20100% of NO_x)
- Gas turbines → yellow/brown plumes

Nitrous Oxide (N_2O)

- Relatively inert
- Uses: Dental anaesthetic
- Strong absorber of infrared radiation (~300 x CO_2)
- Stability = long atmospheric residence times (~150 yrs)
- Hence, potentially significant greenhouse gas
- Long lifetime also allows its transportation into stratosphere and participates in ozone depletion
- Only significant from low-temperature processes (eg. Fluidised bed combustion)

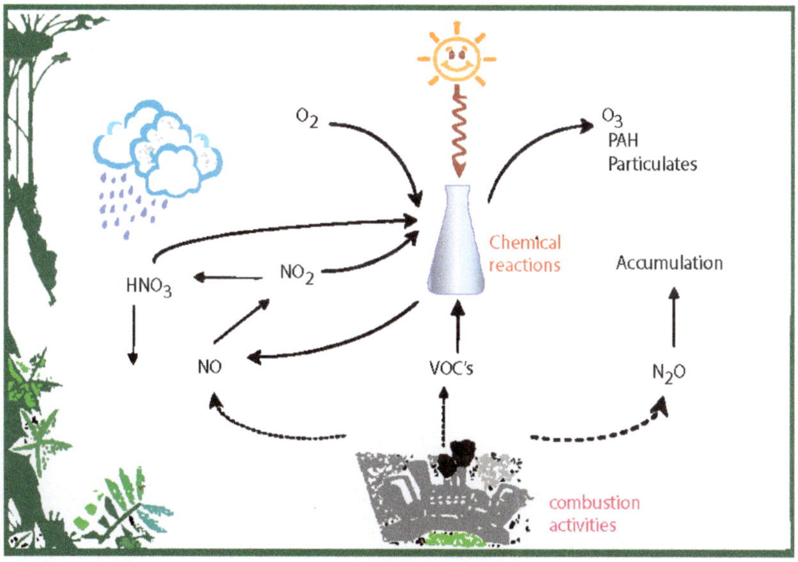

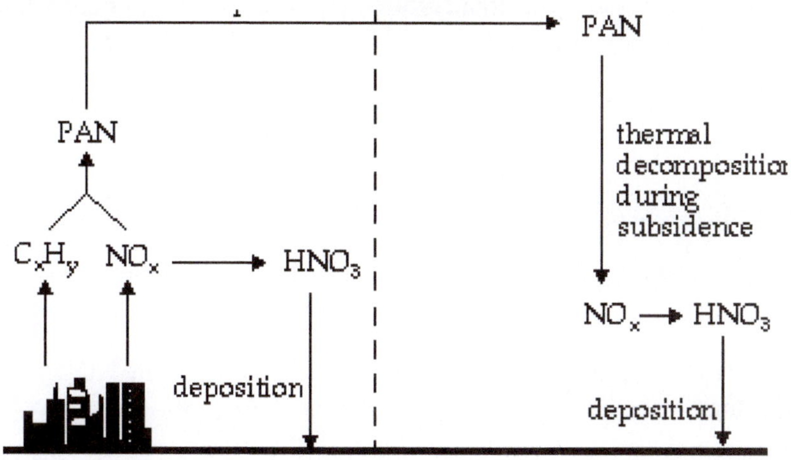

Sources of Nitrogen

- Formation of NO_x requires a source of nitrogen
- Two sources of nitrogen:
  a. Molecular nitrogen from air (1/2 N_2 (from air) +1/2 $O_2 \rightarrow NO$)

     Thermal or Zeldovich Mechanism.

     PromptFenimore Mechanism (HC + N_2).

     Other minor mechanisms

  b. Nitrogen chemically bound within fuel

     Fuel NO_x,
- Most of NO_x in the form of NO

Thermal NO

- One of the most important issues for combustion engineers is: 'What are my NO_x emissions?›
- In most cases, unusually high NO_x emissions are due to NO formed by the Thermal (Zeldovich) mechanism
- Thermal NO mechanism involves the attack of molecular nitrogen (N_2) and atomic nitrogen (N) by oxygen (O_2) and oxygencontaining radicals (O, OH). This can occurs in oxygen rich mixture.
- First identified by Zeldovich (1946) and extended by Fenimore and Jones (1957)
- Described by the following reactions:

  $N_2 + O \rightarrow NO + N$ (R. 1)

  $N + O_2 \rightarrow NO + O$ (R. 2)

  $N + OH \rightarrow NO + H$ (R. 3)
- Westenberg (1971) invoked the steadystate approximation and determined that the maximum NO formation rate is given by:

$$\frac{d[NO]}{dt} = 1.45 \times 10^{17} T^{-1/2} \exp\left[\frac{-69460}{T(K)}\right] . [O_2]_{eq}^{1/2} . [N_2]_{eq} \text{ mol/cm}^3 \text{.s}$$

- Hence, [NO] depends on:
  - Temperature (the higher the temp, strongly the higher the NO formed) ⇒ high temp environment
  - O_2 concentration (the higher the oxygen conc, the higher the NO formed) ⇒ oxygen-rich environment
  - Residence time

Prompt NO

Fenimore (1971) observed an additional formation of NO which could not be explained by the thermal mechanism

- NO formed close to the burner (hence "prompt' NO)
- Effect is not observed under very fuellean conditions or in systems with H_2 or CO as fuel
- Mechanism involves the attack of N_2 by hydrocarbon fuel fragments, mainly CH radicals and Catoms.
- The PromptFenimore mechanism is initiated mainly by R. 4 (the formation of HCN) with a lesser contribution from R. 5:

  $CH + N_2 \rightarrow HCN + N$ (R. 4)

  $C + N_2 \rightarrow CN + N$ (R. 5)

- HCN is subsequently oxidised to NO
- Prompt mechanism dominates for hydrocarbon combustion in fuelrich, in both premixed and diffusion flames

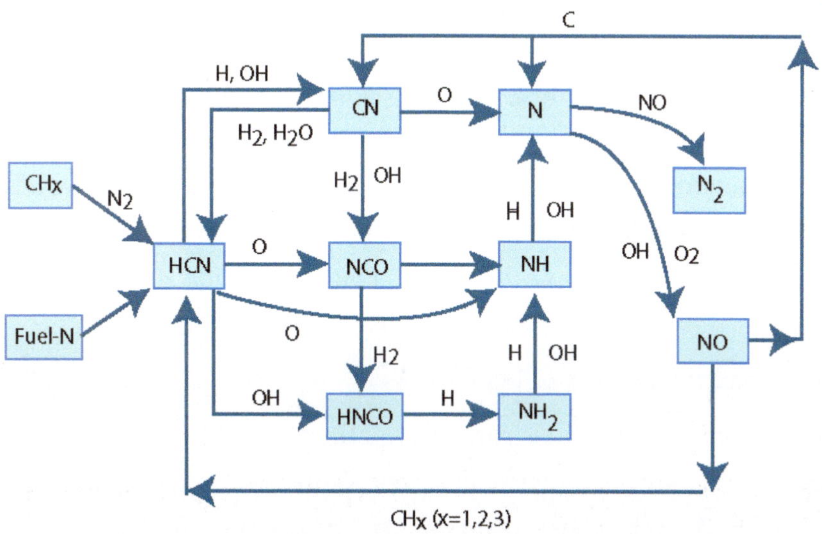

Reaction path diagram illustrating the major steps in **prompt NO formation,** the conversion of volatile-nitrogen in flames, and reburning, (Bowman, 1992)

Minor Mechanisms

1. N_2OIntermediate Mechanism:

- It occurs under fuellean, lowtemperature conditions
- Minor source of NO in most practical combustors
- Mechanism is given by:

$O + N_2 + M \rightarrow N_2O + M$ (R. 6)

$O + N_2O \rightarrow NO + NO$ (R. 7)

$H + N_2O \rightarrow NO + NH$ (R. 8)

2. NNH-Intermediate Mechanism:

- It is observed under laboratory conditions
- H_2 and CH_4 (high H) fuelrich, laminar premixed flames
- Yet to be definitely observed in practical combustors
- Mechanism is given by:

  $H + N_2 \rightarrow NNH$ (R. 9)

  $NNH + O \rightarrow NO + NH$ (R. 10)

Fuel-NO

1. The Nature of FuelNitrogen:

- Nitrogen in coal (and oil) originates in the plant material from which the fuel is formed
- Plants contain nitrogen in the form of proteins, amino acids, alkaloids, chlorophyll and porphyrins
- These were transformed, during the coalification process, into polycyclic aromatic compounds with pyridinic, pyrrolic or other functional groups
- Nitrogen content of coals typically vary between 12.5 wt% and is largely independent of rank

2. Coal Combustion and the Release of FuelNitrogen:

- First stage of coal combustion is rapid devolatilisation of the coal
- Devolatilisation: volatile components such as light hydrocarbons and tars are released and then oxidised in the gas~phase at very short timescales (< 10 ms)
- The solid product of coal pyrolysis is the char which is oxidised at much slower timescales (~ 1s)
- Fuelnitrogen is released during both pyrolysis and char combustion but in very different ways
- Partitioning of fuelnitrogen between the volatiles and char depends on pyrolysis conditions (normally approx. equal)

3. Nitrogen released with the volatiles:

- Nitrogen contained within the volatiles is released as, or rapidly converted to compounds such HCN, NH_3 or HNCO

- These simple nitrogenous species then react in the gas phase to form either NO, N_2O or, under fuelrich conditions to N_2
- Relative amounts of NO, N_2O and N_2 depends strongly on the local O_2 concentration during pyrolysis and temperature
- NO_x formation is reduced if volatiles are released under fuelrich conditions
- Pulverisedfuel combustors NO_x control techniques include: lowNOX burners, flue gas recirculation (fuel-lean) and air staging
- NO emissions are reduced in these techniques by changes in stoichiometry and/or temperature near the burner thus reducing both volatileNO and thermalNO (important to take into account in combustion)

4. Char combustion and the Fate of CharNitrogen:

- The fate of charnitrogen is still leading edge research
- Primary product of charN oxidation is NO
- However, both HCN and HNCO are observed at low temperatures (Ashman et al., 2000)
- NO may be reduced to N_2 by direct reaction with the char surface or by a charcatalysed reaction with CO
- Effective techniques aimed specifically at controlling charNO are not available

5. In principle, NO_x is reduced if it is reacted with C or C_xH_y ⇒ concept of reburn

NO_x Formation

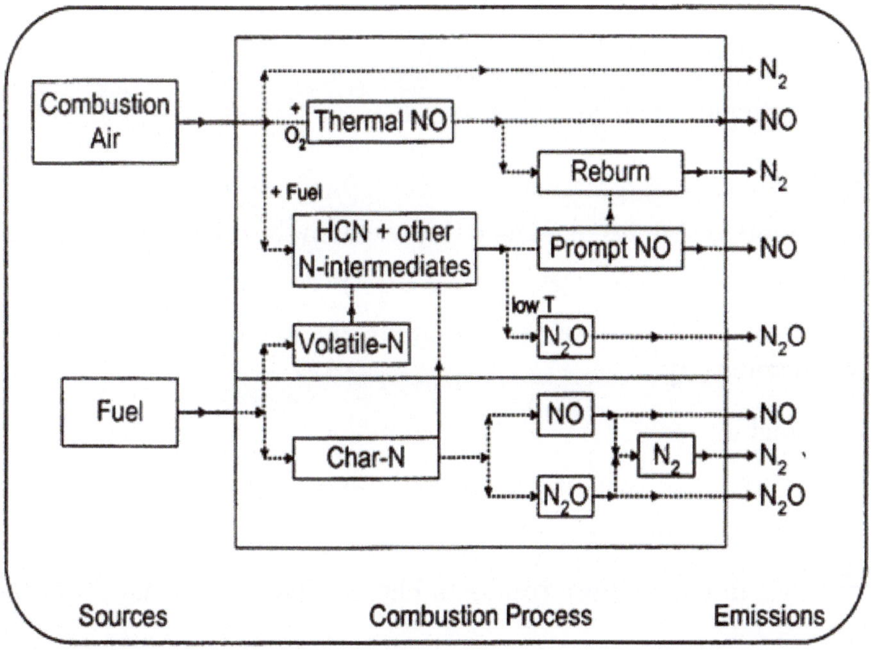

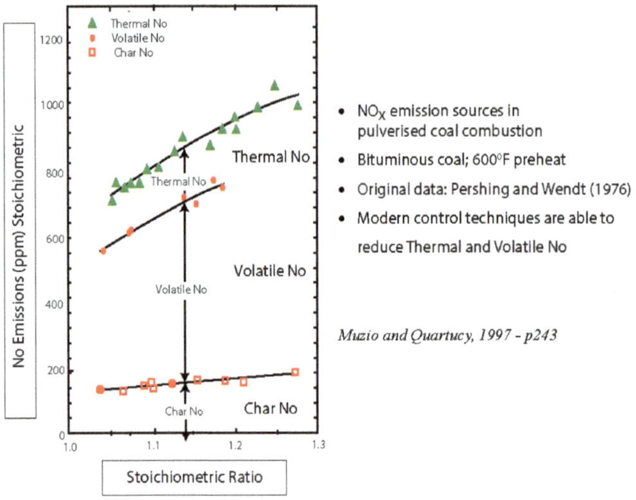

CO (Carbon Monoxide) Formation

- All hydrocarbon fuels are converted to CO_2 via CO
- $CH_4 \rightarrow CH_3 \rightarrow CH_2O \rightarrow CHO \rightarrow CO \rightarrow CO_2$
- Conversion of CO to CO_2 occurs largely downstream of the flamezone and is an equilibrium process which occurs relatively slowly
- Emission of CO as a stable product is minimised if $CO \rightarrow CO_2$ is allowed to proceed to completion
- There are 3 processes leading to incomplete combustion:
    - o Insufficient Oxygen (low O_2)
    - o Aerodynamic Quenching (low temp)
    - o Impingement Quenching (low temp)

1. Insufficient Oxygen

- In this case, there is simply not enough O_2 present in the downstream gases to fully oxidise the CO to CO_2
- Usually as a result of poor design or operation and, in flames, often related to inadequate mixing of secondary air

2. Aerodynamic Quenching

- Caused by a rapid decrease in the temperature of the postflame gases, either by expansion of the gases or by mixing with ambient air
- Rapid decrease in temperature results in kinetic "freezing" of the CO at the equilibrium condition

- Aerodynamic quenching is a serious problern for spark-ignition engine when too much air is used (not really for openflame systems)

3. Impingement Quenching

- Quenching occurs as a result of the hot postflame gases impinging on a relatively cold surface

- Again, a problem in internal combustion engines

CO Toxicity

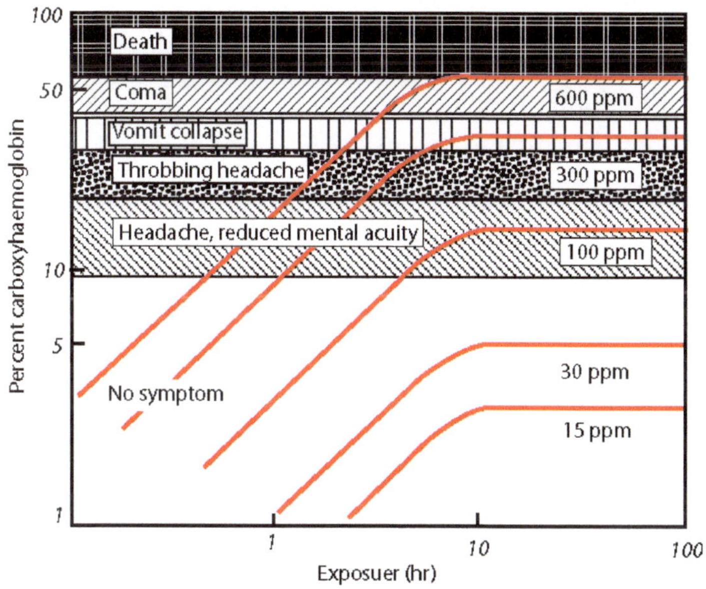

Turns, 1st Edition, p474

Balance Between CO and NO

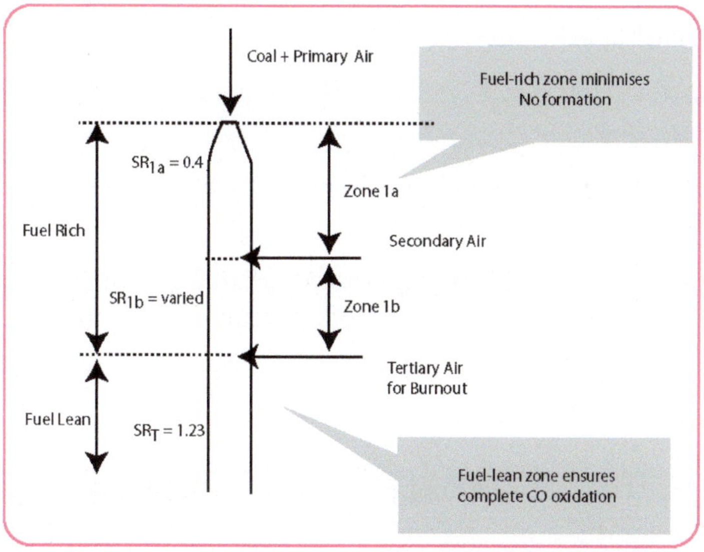

- We have seen that CO is largely a product of insufficient combustion (low O_2, rapid cooling, etc)
- In comparison, the formation of NO which is enhanced by conditions which could be termed as overaggressive combustion (high O_2, high temp, etc)
- There is a fine balance between the formation of NO and CO such that control measures which are effective against one species often result in an increase in the other species
- Fortunately, the two mechanisms are decoupled somewhat since NO formation occurs primarily in the nearburner zone and CO consumption occurs in the postflame gases.

Volatile Organic Compounds (VOC's)

- Volatile organic compounds (VOC's) are also products of incomplete combustion
- VOC consist of various unburned fuel fragments, aliphatic and aromatic hydrocarbons and partially oxidised fuel fragments
- The formation and destruction of VOC's occurs much closer to the flame front (higher temp zone) than does CO and hence greater extents of incomplete combustion cause the emission of VOC's
- VOC's play a key role in photochemical smog cycles (reactions in atmosphere triggered by photon of the sunray)

PAH's, Soot & PM10 (solid phase pollutants)

What are PAH's, Soot & PM10?

- PAH's (polycyclic aromatic hydrocarbons) are clusters of aromatic rings up to about 4 rings in size.
- Soot is formed when, under appropriate conditions, PAH's grow in size until they reach approximately 2050 nm
- PM10 is a general term given to all airborne particulates which have a size < 10 microns

PAH's (polycyclic aromatic hydrocarbons):

- PAH's are formed in, and may be emitted from, combustion processes under very fuelrich conditions
- The main cause of concern with PAH's is that many species are known mutagens, cocarcinogens, or carcinogens
- While PAH's are oxidised quite rapidly in the urban atmosphere, they may be stabilised considerably by adsorption to the surface of particulate (e.g. soot)
- PAH's then become quite long-lived and are capable of penetrating deep into the human lungs

Soot:

- The health effects of soot are serious and are related to the transport of fine particles within the lungs

- PM10 is the environmental standard used to quantify airborne particulates from all sources
- Soot particles also contribute to visible pollution and haze within the urban environment
- Soot formation is encouraged within some combustion applications on the basis of the significantly enhanced radiative heat transfer obtained from sooting flames

PAH & Soot Formation:

- PAH's and soot are formed mainly in diffusion flames and in diesel engines (also from I.C. engines with lubricant oil leakage)
- A precursor species (believed to be acetylene) is formed under fuelrich conditions
- Acetylene and other hydrocarbon fragments combine through pyrolytic (without oxygen) reactions to form progressively larger ring structures (PAH's)
- Without competition with oxidation reactions, which break ring structures, the clusters continue to increase in size
- Small particles of a critical size are formed by the growth of these aromatic clusters by both chemical and physical means, i.e. coagulation
- At this point, the molecules are identifiable as primary soot particles
- The soot particles continue to grow by both chemical (surface growth) and physical (particle agglomeration) processes whilst they remain within the bath of precursor species, clusters and adolescent soot particles

Sulphur Compounds

- All of the sulphur contained within various fuels is emitted from combustion processes as either SO_2 or SO_3
- The ultimate fate of oxides of sulphur in the environment is transformed to H_2SO_4 (acid rain)
- Only two possible abatement procedures:
- remove S before combustion (from the fuel)
- remove SO_x after combustion (from the flue gases)
- Most common technique is the reaction of SO_2 in the flue gas with limestone ($CaCO_3$) or lime (CaO)
- A slurry of limestone or lime is sprayed in an aqueous absorption tower through which the flue gas is passed

Coal Combustion

- What is coal? Formation, sources, applications.

- Coal combustion description.
- Coal power plants and air pollution: Mechanisms and control technologies.
- Coal and air pollution in Israel.

Coal – what is it?

Elemental Composition	C	65-95%
	H	2-7%
	O	<25%
	S	<10%
	N	1-2%
Proximate Analysis	Char	20-70%
	Ash	5-15%
	H_2O	2-20%
	VM	20-45%

- Inhomogeneous organic fuel formed mainly from decomposed plant matter.
- Over 1200 coals have been classified.
- Coalification forms different coal types

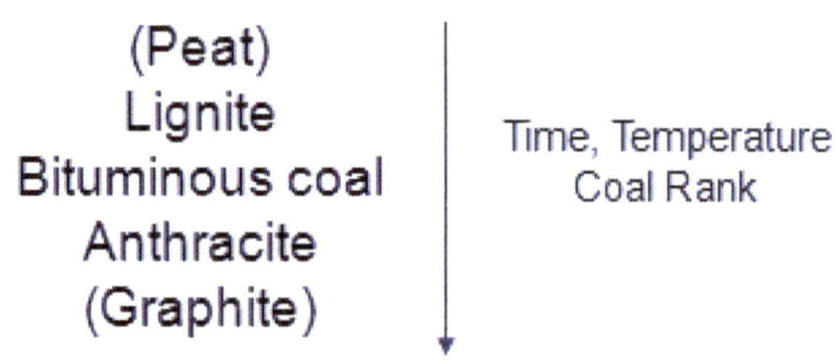

Main Processes in Coal Combustion

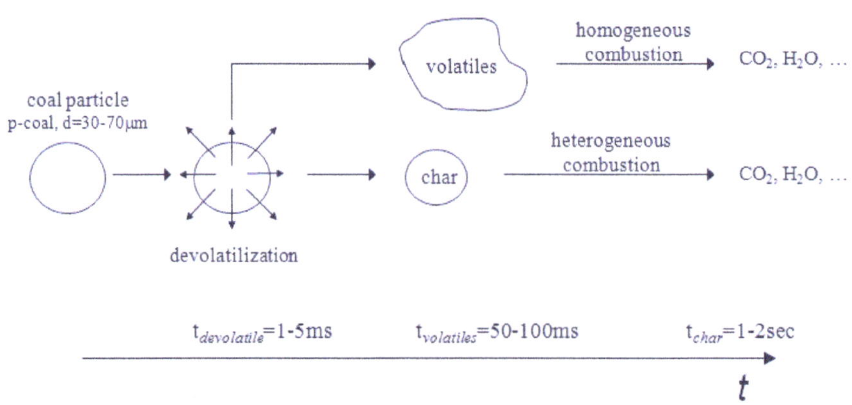

The physical processes influencing pulverized coal combustion

- Turbulent/swirling flow of air and coal.
- Turbulent/convective/molecular diffusion of gaseous reactants and products.

- Convective heat transfer through the gas and between the gas and coal particles.
- Radiative heat transfer between the gas and coal particles and between the coal/air mixture and the furnace walls.

From Fumifugium by John Evelyn (1661) - on London's air pollution -

"...but so universally mixed with the otherwise wholesome and excellent, that her Inhabitants breathe nothing but an impure and thick Mist, accompanied by a fuliginous and filthy vapour, which renders them obnoxious to a thousand inconveniences, corrupting the Lungs, and disordering the entire habit of their Bodies; so that Catharrs, Phthisicks, Coughs and Consumptions, rage more in this one City, than the whole Earth besides.For when in all other places the Aer is most Serene and Pure, it is here Ecclipsed with such a Cloud of Sulphure, as the Sun itself, which gives day to all the World besides, is hardly able to penetrate and impart it here; and the weary Traveller, at many Miles distance, sooner smells, than sees the City to which he repairs. This is that pernicious Smoake which sullyes all her Glory, superinducing a sooty Crust or Fur upon all that it lights, spoyling the moveables, tarnishing the Plate, Gildings and Furniture, and corroding the very Iron-bars and hardest Stones with those piercing and acrimonious Spirits which accompany its Sulphure; and executing more in one year, than exposed to the pure Aer of the Country it could effect in some hundreds."

Coal Combustion Air Pollutants

• CO_2 • CO • NO_x • SO_x • Particulate matter • Trace metals • Organic compounds	 

Carbon Dioxide, CO_2

$$C + O_2 \longrightarrow CO_2$$

Almost 99% of C in coal is converted to CO_2.In order to lower CO_2 emission levels, coal power plants will have to leave steam-based systems (37% efficiency) and go towards coal gasification technology (60% efficiency).Meanwhile, CO_2 sequestration is being tested.

Carbon Monoxide, CO

$$C + \tfrac{1}{2}O_2 \longrightarrow CO$$

CO is minimized by control of the combustion process (air/fuel ratio, residence time, temperature or turbulence).

Particulate Matter

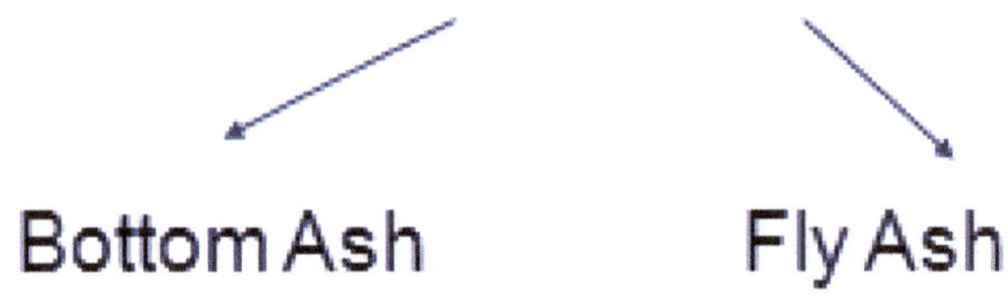

PM composition and emission levels are a complex function of:

1. Coal properties,
2. Boiler firing configuration,
3. Boiler operation,
4. Pollution control equipment.

In PC power plants, since combustion is almost complete, the emitted PM is primarily composed of inorganic ash residues.

PM Controls (AP-42, EPA)

Mainly post combustion methods:

Electrostatic precipitator (ESP)	99% (for 0.1>d(μm)>10) <99% (for 0.1<d (μm)<10)
Fabric filter (or baghouse)	As high as 99.9%
Wet scrubber	95-99%
Cyclone	90-95% (d(μm)>10)

Trace Metals

FORMATION

Concentration of metal in coal, physical and chemical properties of the metal, combustion conditions.

Class 1
Elements that are approximately equally concentrated in the fly ash and bottom ash (Mn, Be, Co, Cr)

Class 2
Elements that are enriched in fly ash relative to bottom ash (Ar, Cd, Pb, An)

Class 3
Elements which are emitted in the gas phase (mainly Hg).

CONTROL

Control of total particulate matter emissions

Collection of fine particles.

Sorbents ???

Organic Compounds

Include volatile, semivolatile and condensable organic compounds either present in the coal or formed as a product of incomplete combustion.

Characterized by hydrocarbon class: alkanes, alkenes, aldehydes, alcohols and substituted benzenes.

The main groups of environmental concern are:

1) tetrachloro- through octachloro- dioxins and furnans.

2) Polycyclic organic matter (POM).

Emissions dependent on combustion behavior in the boiler (air/fuel ratio, residence time, temperature or turbulence).

Sulfur Oxides, SO_x

Sulfur in coal (<10%)

Organic sulfur (40%)

Chemically bonded to the hydrocarbon matrix in the forms of thiophene, thiopyrone, sulfides and thiol.

Inorganic sulfur (60%)

Imbedded in the coal, as loose pyrite - FeS_2 or marcasite, and calcium/iron/barium sulfates.

Sources of sulfur in coal: Seawater sulfates, Limestone

SOx Formation

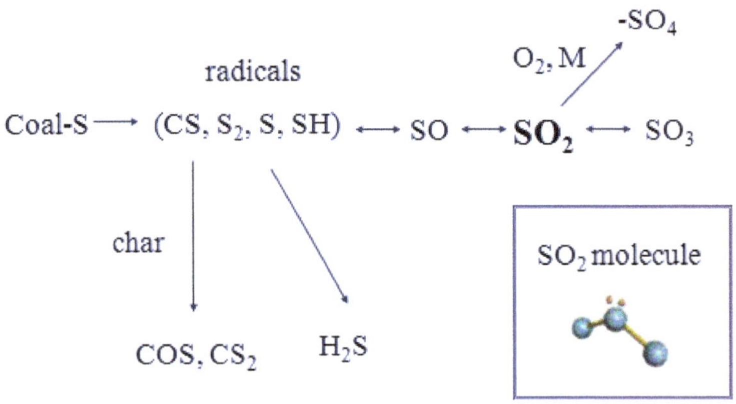

Nitrogen Oxides, NO_x

Nitrogen in Coal (1-2%)

Name	Structure	~ Relative amount	Stability
Pyridine[1]	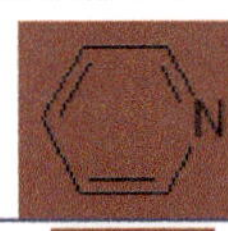	15-40%	More stable
Pyrrole[1]		60%	Less stable
Aromatic amines	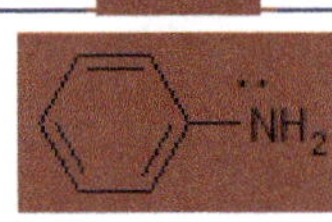	6-10%	Stable

Including structures made up of 2-5 fused aromatic rings.

Main NO Mechanisms

1.Thermal NO

2.Prompt NO

3.Fuel NO: volatiles-NO and char-NO

Thermal NO

(Zeldovich mechanism)

$$N_2 + O \leftrightarrow NO + N$$
$$N + O_2 \leftrightarrow NO + O$$

Strong temperature-dependence: >1300-1500°C

Not a major source of NO in coal utility boilers.

Prompt NO

$$N_2 + CH_x \leftrightarrow HCN + N + \ldots$$
$$N + OH \leftrightarrow NO + H$$

Prevalent only in fuel-rich systems.

Not a major source of NO in coal utility boilers.

Fuel NO (-N in volatiles)

$$\text{Fuel-N} \xrightarrow{\text{volatiles}} \mathbf{HCN}/NH_3$$

$$HCN/NH_3 + O_2 \longrightarrow NO \quad \text{(formation)}$$
$$NO + HCN/NH_3 \longrightarrow N_2 \quad \text{(destruction)}$$

The major source of NO in coal utility boilers (>80%).

Char NO (-N in the char)

$$\text{Char-N} + \tfrac{1}{2}O_2 \rightarrow NO \quad \text{(formation)}$$
$$\text{Char-C} + NO \rightarrow \tfrac{1}{2}N_2 + \text{Char(O)} \quad \text{(destruction)}$$

$$[\text{char-NO} = \sim 25\%] < [\text{volatiles-NO} = \sim 75\%]$$

Photochemical Smog

- Photochemical smog is a mixture of air pollutants formed by the reaction of nitrogen oxides and volatile organic hydrocarbons when they are exposed to sunlight.

- More than 100 primary and secondary pollutants can be formed.

VOC + NO_x + heat + sunlight ⇨ ground level ozone (O_3) ++ other photochemical oxidants + aldehydes + other secondary air pollutants

Temperature Inversions

During daylight hours, the sun warms the ground which warms the air near the Earth's surface. The warm air rises as more dense cool air displaces it. Any pollution in the air normally disperses in the turbulence caused by the mixing of the air. When a layer of cool air traps the layer of warm air near the Earth's surface, then the situation is known as a temperature inversion. In this case, pollutants are not dispersed but continue to build up in the still, stagnant air.

Subsidence Inversion

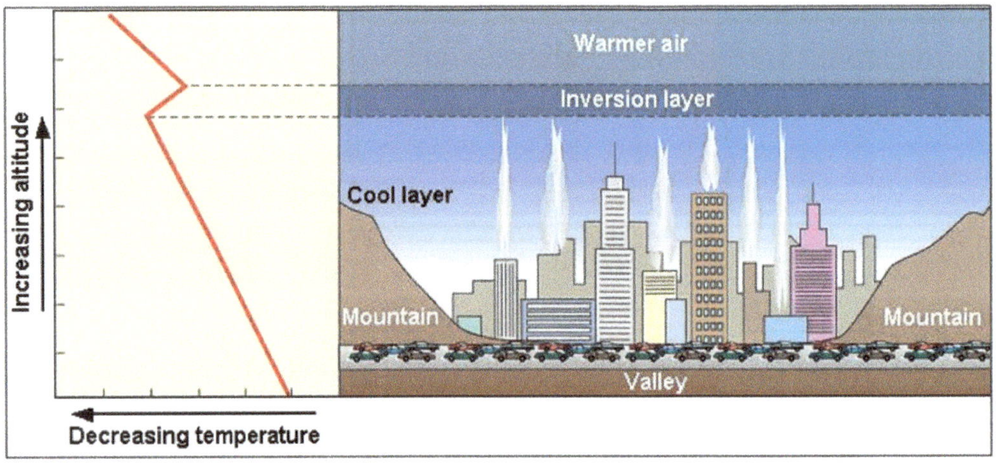

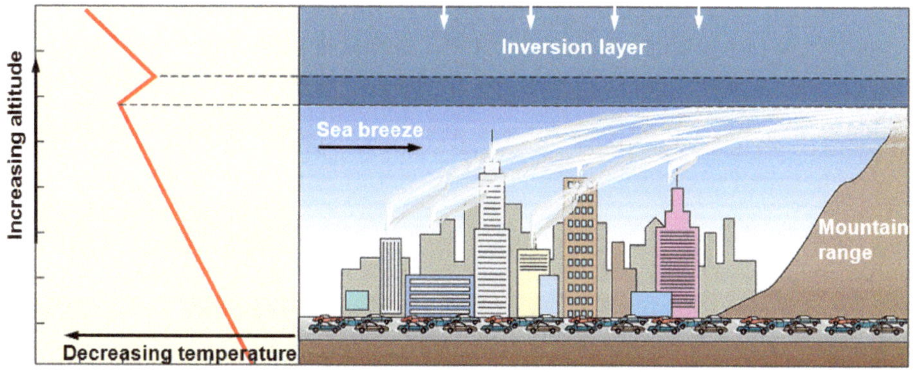

Sea breezes keep the air and pollution onshore, and the mountains prevent the air from escaping. The cold air aloft acts as a lid.

Photochemical Smog

Brown Air Smog

Photochemical

Reaction

Photochemical

oxidents

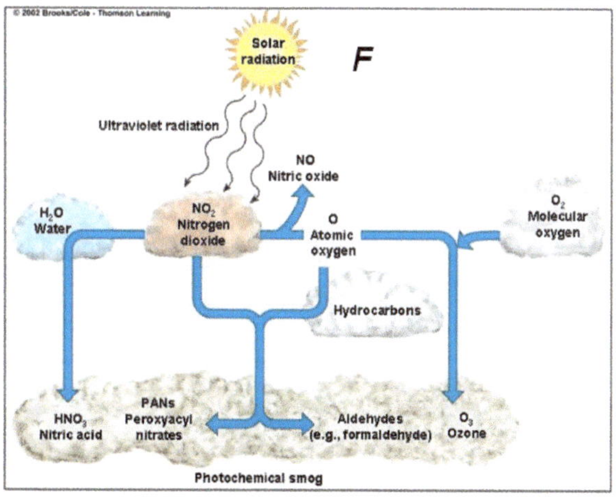

Industrial Smog

Particulates

Sulfur Dioxide

Sulfuric Acid

Gray Air Smog

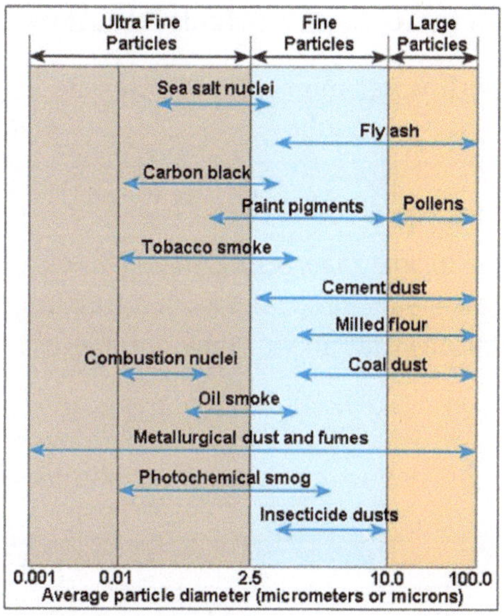

The oxygen atom reacts with other components of automobile exhaust (e.g. unburnt hydrocarbons) and those of the atmosphere (e.g. oxygen) in a series of complex reactions to produce a variety of lachrymatory and toxic chemicals (e.g. peroxyacetyl nitrate).

The production and destruction of photochemical smog are very complex events that involve hundreds of different chemical reactions.

In the first step of photochemical smog production, sunlight acts on nitrogen dioxide to produce nitric oxide and free oxygen:

$$NO_2 + hv \rightarrow NO + O$$

Oxygen atoms formed in this reaction react readily with diatomic oxygen present in the atmosphere to form ozone (O_3). The symbol *M* in the following equation represents some third body that acts as a catalyst, removing energy from the O_2/O collision, making it thermodynamically feasible:

$$O + O_2 + M \rightarrow O_3 + M$$

Levels of ozone produced by this reaction are controlled to some extent by a "scavenging" reaction that takes place with nitric oxide:

$$O_3 + NO \rightarrow O_2 + NO_2$$

The reaction is referred to by this term because nitric oxide forages among ozone molecules, attacking them at random and converting them to oxygen molecules.

These three reactions constitute a cycle that begins with the production of nitrogen dioxide (usually from motor vehicle exhaust) and results in the production of large amounts of atomic oxygen and ozone.

When concentrations of nitrogen dioxide decrease (as during evening hours, when traffic tends to decrease), the scavenging reaction tends to remove excess amounts of ozone from the atmosphere.

Oxygen atoms formed in this cycle of reactions initiate one of two important chain reactions.

In the first, they react with water molecules to generate highly reactive (indicated by the asterisk) hydroxyl radicals:

$$O^* + H_2O \rightarrow 2OH^*$$

In the second, the oxygen atoms react with hydrocarbons (VOCs) present in automotive exhaust to produce *free radicals* (atoms or molecules with a free unpaired electron, indicated by the symbol R• in the following equation) that may or may not contain oxygen:

$$O + RH \rightarrow R\cdot$$

Reactive hydroxyl radicals and ozone may initiate similar changes:

$$OH^* + RH \rightarrow R\cdot + H_2O$$

$$O_3 + RH \rightarrow R\cdot$$

The free radical R• is especially important because it reacts very readily with oxygen to form the *peroxyl radical*, ROO•, which may itself initiate a series of important reactions:

$$R^{\cdot} + O_2 \rightarrow ROO^{\cdot}$$

One of the most important of these reactions occurs between a peroxyl radical and nitrogen dioxide to produce compounds known as peroxyalkyl and peroxyacyl nitrates:

$$ROO^{\cdot} + NO_2 \rightarrow ROONO_2$$

One product commonly formed in this reaction is peroxyacetyl nitrate (PAN), an important and common component of photochemical smog.

PAN, ozone, and many of the other compounds formed in the series of reactions just described are strong oxidants and are responsible for some of the initial and most aggravating features of smog, including nose, eye, and throat irritation.

References

- Staff (12 November 2015). "Audio (66:01) - NASA News Conference - Carbon & Climate Telecon". NASA. Retrieved 12 November 2015
- J. Sunyer (2001). "Urban air pollution and Chronic Obstructive Pulmonary disease: a review". European Respiratory Journal. 17 (5): 1024–1033. doi:10.1183/09031936.01.17510240. PMID 11488305.
- Davis, Devra (2002). When Smoke Ran Like Water: Tales of Environmental Deception and the Battle Against Pollution. Basic Books. ISBN 0-465-01521-2
- "NATA | National-Scale Air Toxics Assessments | Technology Transfer Network Air Technical Web Site | US EPA". Epa.gov. 2006-06-28. Retrieved 2012-12-11
- J. C. Fensterstock , J. A. Kurtzweg & G. Ozolins (1971): Reduction of Air Pollution Potential through Environmental Planning, Journal of the Air Pollution Control Association, 21:7, 395-399
- Willey, Joanne M. (2011). Prescott's Microbiology 8th Ed. New York, N.Y.: McGraw Hill. p. 705. ISBN 978-0-07-337526-7
- European Court of Justice, CURIA (2008). "PRESS RELEASE No 58/08 Judgment of the Court of Justice in Case C-237/07" (PDF). Retrieved 24 January 2015
- European Commission. "Air quality: Commission sends final warning to UK over levels of fine particle pollution". Archived from the original on 11 May 2011. Retrieved 7 April 2011.
- Miller, Charles (2008). Biological Oceanography. 350 Main Street, Malden, MA 02148 USA: Blackwell Publishing Ltd. pp. 60–62. ISBN 978-0-632-05536-4
- House of Commons Environmental Audit Committee (2010). "Environmental Audit Committee - Fifth Report Air Quality". Retrieved 24 January 2015
- Marchant, H. K., Lavik, G., Holtappels, M., and Kuypers, M. M. M. (2014). "The Fate of Nitrate in Intertidal Permeable Sediments". PLoS ONE. 9 (8). doi:10.1371/journal.pone.0104517
- Gruber, Nicolas (2008). Nitrogen in the Marine Environment. 30 Corporate Drive, Suite 400, Burlington, MA 01803: Elsevier. pp. 1–35. ISBN 978-0-12-372522-6
- Guardian (11 March 2011). "Britain fends off threat of £300m fine over London air pollution". Guardian. Retrieved 24 January 2015
- Committee on Environmental Health (2004). "Ambient Air Pollution: Health Hazards to Children". Pediatrics. 114 (6): 1699–1707. PMID 15574638. doi:10.1542/peds.2004-2166

- BBC (2010) (2010-12-06). "Threat to sue over London congestion charge scrapping". BBC News. Retrieved 24 January 2015
- Lalli, Parsons, Carol, Timothy (1997). Biological oceanography: An introduction. Butterworth-Heinemann. ISBN 978-0-7506-3384-0
- "Many Planets, One Earth // Section 4: Carbon Cycling and Earth's Climate". Many Planets, One Earth. 4. Retrieved 2012-06-24
- Pope, CA; Aruni Bhatnagar; James P. McCracken; Wesley T. Abplanalp; Daniel J. Conklin; Timothy E. O'Toole (2016). "Exposure to Fine Particulate Air Pollution Is Associated with Endothelial Injury and Systemic Inflammation". Circulation Research. 119: 1204–1214. PMID 27780829. doi:10.1161/circresaha.116.309279
- Buis, Alan; Ramsayer, Kate; Rasmussen, Carol (12 November 2015). "A Breathing Planet, Off Balance". NASA. Retrieved 13 November 2015
- Chapin, S. F. III, Matson, P. A., Mooney H. A. 2002. Principles of Terrestrial Ecosystem Ecology. Springer, New York 2002 ISBN 0-387-95443-0, p.345
- St. Fleur, Nicholas (10 November 2015). "Atmospheric Greenhouse Gas Levels Hit Record, Report Says". New York Times. Retrieved 11 November 2015
- Crowley, T. J. (2000). "Causes of Climate Change Over the Past 1000 Years". Science. 289 (5477): 270–277. Bibcode:2000Sci...289..270C. ISSN 0036-8075. PMID 10894770. doi:10.1126/science.289.5477.270
- Ritter, Karl (9 November 2015). "UK: In 1st, global temps average could be 1 degree C higher". AP News. Retrieved 11 November 2015

3

Water Pollution and Aquatic Toxicology

When water bodies are contaminated with substances which are harmful for biotic or non-biotic components of the Earth, it is referred to as water pollution. This degradation is studied in the context of aquatic toxicology, which deals with effects of chemicals on aquatic bodies. This chapter studies in detail water pollution and aquatic toxicology.

Water Pollution

Any chemical, biological, physical change in water quality that has a harmful effect on living organisms or makes water unsuitable for desired usage.

WHO:

- 3.4 million premature deaths each year from waterborne diseases
- 1.9 million from diarrhea

- U.S. 1.5 million illnesses
- 1993 Milwaukee 370,000 sick

Infectious Agents: bacteria and viruses often from animal wastes

Oxygen Demanding Wastes: organic waste that needs oxygen often from animal waste, paper mills and food processing.

Inorganic Chemicals: Acids and toxic chemicals often from runoff, industries and household cleaners

Organic Chemicals: oil, gasoline, plastics, detergents often from surface runoff, industries and cleaners

Plant Nutrients: water soluble nitrates, ammonia and phosphates often from sewage, agriculture and urban fertilizers

Sediment: soils and silts from land erosion can disrupt photosynthesis, destroy spawning grounds, clog rivers and streams

Heat Pollution and Radioactivity: mostly from powerplants

How do We Measure Water Quality

Bacterial Counts: Fecal coliform counts from intestines of animals

- None per 100 ml for drinking
- >200 per 100 ml for swimming

Sources: human sewage, animals, birds, raccoons, etc.

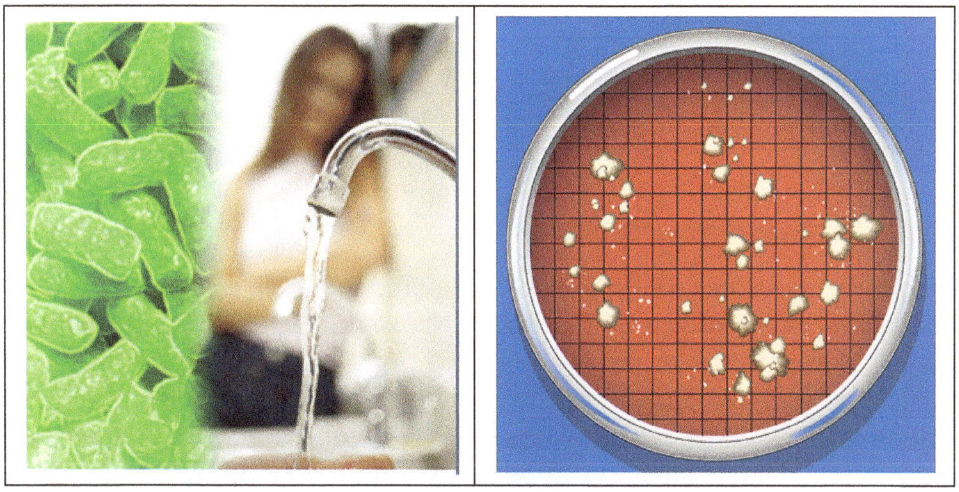

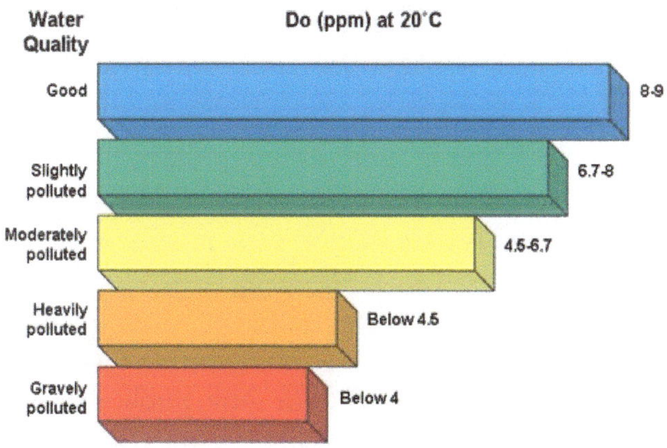

Dissolved Oxygen: BOD Biological Oxygen Demand...the amount of oxygen consumed by aquatic decomposers

Chemical Analysis: looking for presence of inorganic or organic chemicals

Suspended Sediment water clarity

Indicator Species: organisms that give an idea of the health of the water body.

- Mussels, oysters and clams filter water

Aquatic Toxicology

Aquatic toxicology is the study of the effects of manufactured chemicals and other anthropogenic and natural materials and activities on aquatic organisms at various levels of organization, from subcellular through individual organisms to communities and ecosystems. Aquatic toxicology is a multidisciplinary field which integrates toxicology, aquatic ecology and aquatic chemistry.

This field of study includes freshwater, marine water and sediment environments. Common tests include standardized acute and chronic toxicity tests lasting 24–96 hours (acute test) to 7 days or more (chronic tests). These tests measure endpoints such as survival, growth, reproduction, that are measured at each concentration in a gradient, along with a control test. Typically using selected organisms with ecologically relevant sensitivity to toxicants and a well-established literature background. These organisms can be easily acquired or cultured in lab and are easy to handle.

History

While basic research in toxicology began in multiple countries in the 1800s, it was not until around the 1930s that the use of acute toxicity testing, especially on fish, was established. Over the next two decades, the effects of chemicals and wastes on non-human species became more of a public issue and the era of the *pickle-jar bioassays* began as efforts increased to standardize toxicity testing techniques. In the United States of America, the passage of the Federal Water Pollution Control Act of 1947 marked the first comprehensive legislation for the control of water pollution and was followed by the Federal Water Pollution Control Act in 1956. In 1962, public and governmental interests were renewed, in large part due to the publication of Rachel Carson's *Silent Spring*, and three years later the Water Quality Act was passed which directed states to develop water quality standards. Public awareness, as well as scientific and governmental concern, continued to grow throughout the 1970s and by the end of the decade research had expanded to include hazard evaluation and risk analysis. In the subsequent decades, aquatic toxicology has continued to expand and internationalize so that there is now a strong application of toxicity testing for environmental protection.

Aquatic Toxicity Tests

Aquatic toxicology tests (assays): toxicity tests are used to provide qualitative and quantitative data on adverse (deleterious) effects on aquatic organisms from a toxicant. Toxicity tests can be used to assess the potential for damage to an aquatic environment and provide a database that can be used to assess the risk associated within a situation for a specific toxicant. Aquatic toxicology tests can be performed in the field or in the laboratory. Field experiments generally refer to multiple species exposure and laboratory experiments generally refer to single species exposure. A dose response relationship is most commonly used with a sigmoidal curve to quantify the toxic effects at a selected end-point or criteria for effect (i.e. death or other adverse effect to the organism). Concentration is on the x-axis and percent inhibition or response is on the y-axis.

The criteria for effects, or endpoints tested for, can include lethal and sublethal effects.

There are different types of toxicity tests that can be performed on various test species. Different species differ in their susceptibility to chemicals, most likely due to differences in accessibility, metabolic rate, excretion rate, genetic factors, dieteary factors, age, sex, health and stress level of the organism. Common standard test species are the fathead minnow (Pimephales promelas), daphnids (*Daphnia magna, D. pulex, D. pulicaria, Ceriodaphnia dubia*), midge (Chironomus tentans, C. ruparius), rainbow trout (Oncorhynchus mykiss), sheepshead minnow (Cyprinodon variegatu), mysids (Mysidopsis), oyster (Crassotreas), scud (Hyalalla Azteca), grass shrimp (Palaemonetes pugio), mussels (Mytilus). As defined by ASTM, these species are routinely selected on the basis of availability, commercial, recreational, and ecological importance, past successful use, and regulatory use.

A variety of acceptable standardized test methods have been published. Some of the more widely accepted agencies to publish methods are: the American Public Health Association, U.S. Environmental Protection Agency, American Society for Testing and Materials, International Organization for Standardization, Environment Canada, and Organization for Economic Cooperation and Development. Standardized tests offer the ability to compare results between laboratories.

There are many kinds of toxicity tests widely accepted in the scientific literature and regulatory agencies. The type of test used depends on many factors: Specific regulatory agency conducting the test, resources available, physical and chemical characteristics of the environment, type of toxicant, test species available, laboratory vs. field testing, end-point selection, and time and resources available to conduct the assays are some of the most common influencing factors on test design.

Exposure Systems

Exposure systems are four general techniques the controls and test organisms are exposed to the dealing with treated and diluted water or the test solutions.

Static- a static test exposes the organism in still water. The toxicant is added to the water in order to obtain the correct concentrations to be tested. The control and test organisms are placed in the test solutions and the water is not changed for the entirety of the test.

Recirculation- a recirculation test exposes the organism to the toxicant in a similar manner as the static test, except that the test solutions are pumped through an apparatus (i.e. filter) to maintain water quality, but not reduce the concentration of the toxicant in the water. The water is circulated through the test chamber continuously, similar to an aerated fish tank. This type of test is expensive and it is unclear whether or not the filter or aerator has an effect on the toxicant.

Renewal- a renewal test also exposes the organism to the toxicant in a similar manner as the static test because it is in still water. However, in a renewal test the test solution is renewed periodically (constant intervals) by transferring the organism to a fresh test chamber with the same concentration of toxicant.

Flow-through- a flow through test exposes the organism to the toxicant with a flow into the test chambers and then out of the test chambers. The once-through flow can either be intermittent or continuous. A stock solution of the correct concentrations of contaminant must be previously prepared. Metering pumps or diluters will control the flow and the volume of the test solution, and the proper proportions of water and contaminant will be mixed.

Types of Tests

Acute tests are short-term exposure tests (hours or days) and generally use lethality as an end-point. In acute exposures, organisms come into contact with higher doses of the toxicant in a single event or in multiple events over a short period of time and usually produce immediate effects, depending on absorption time of the toxicant. These tests are generally conducted on organisms during a specific time period of the organism's life cycle, and are considered partial life cycle tests. Acute tests are not valid if mortality in the control sample is greater than 10%. Results are reported in EC50, or concentration that will affect fifty percent of the sample size.

Chronic tests are long-term tests (weeks, months years), relative to the test organism's life span (>10% of life span), and generally use sub-lethal endpoints. In chronic exposures, organisms come into contact with low, continuous doses of a toxicant. Chronic exposures may induce effects to acute exposure, but can also result in effects that develop slowly. Chronic tests are generally considered full life cycle tests and cover an entire generation time or reproductive life cycle ("egg to egg"). Chronic tests are not considered valid if mortality in the control sample is greater than 20%. These results are generally reported in NOECs (No observed effects level) and LOECs (Lowest observed effects level).

Early life stage tests are considered as subchronic exposures that are less than a complete reproductive life cycle and include exposure during early, sensitive life stages of an organism. These exposures are also called critical life stage, embryo-larval, or egg-fry tests. Early life stage tests are not considered valid if mortality in the control sample is greater than 30%.

Short-term sublethal tests are used to evaluate the toxicity of effluents to aquatic organisms. These methods are developed by the EPA, and only focus on the most sensitive life stages. Endpoints for these test include changes in growth, reproduction and survival. NOECs, LOECs and EC50s are reported in these tests.

Bioaccumulation tests are toxicity tests that can be used for hydrophobic chemicals that may accumulated in the fatty tissue of aquatic organisms. Toxicants with low solubilities in water generally can be stored in the fatty tissue due to the high lipid content in this tissue. The storage of these toxicants within the organism may lead to cumulative toxicity. Bioaccumulation tests use bioconcentration factors (BCF) to predict concentrations of hydrophobic contaminants in organisms. The BCF is the ratio of the average concentration of test chemical accumulated in the tissue of the test organism (under steady state conditions) to the average measured concentration in the water.

Freshwater tests and saltwater tests have different standard methods, especially as set by the regulatory agencies. However, these tests generally include a control (negative and/or positive), a geometric dilution series or other appropriate logarithmic dilution series, test chambers and equal numbers of replicates, and a test organism. Exact exposure time and test duration will depend on type of test (acute vs. chronic) and organism type. Temperature, water quality parameters and light will depend on regulator requirements and organism type.

Effluent toxicity tests are tests conducted under the Clean Water Act, National Pollutant Discharge Elimination System (NPDES) permit program and are used by dischargers of contaminated effluent to monitor the quality of effluent into receiving waters. Acute Effluent Toxicity Tests are used to monitor the quality of industrial effluent monthly using acute toxicity tests. Effluent is used to perform static-acute multi concentration toxicity tests with *Ceriodaphnia dubia* and *Pimephales promelas*. The test organisms are exposed for 48 hours under static conditions with five concentrations of the effluent. Short-term Chronic Effluent Toxicity Tests are used to monitor the quality of municipal wastewater treatment plants effluent quarterly using short-term chronic toxicity tests. The goal of this test is to ensure that the wastewater is not chronically toxic. The major deviation in the short-term chronic effluent toxicity tests and the acute effluent toxicity tests is that the short-term chronic test lasts for seven days and the acute test lasts for 48 hours.

Sediment Tests

At some point most chemicals originating from both anthropogenic and natural sources accumulate in sediment. For this reason, sediment toxicity can play a major role in the adverse biological effects seen in aquatic organisms, especially those inhabiting benthic habitats. A recommended approach for sediment testing is to apply the Sediment Quality Triad (SQT) which involves simultaneously examining sediment chemistry, toxicity, and field alterations so that more complete information can be gathered. Collection, handling, and storage of sediment can have an effect on bioavailability and for this reason standard methods have been developed to suit this purpose.

Toxicological Effects

Toxicity can be broken down into two broad categories of direct and indirect toxicity. Direct toxicity results from a toxicant acting at the site of action in or on the organism. Indirect toxicity occurs with a change in the physical, chemical, or biological environment.

Lethality is most common effect used in toxicology and used as an endpoint for acute toxicity tests. While conducting chronic toxicity tests sublethal effects are endpoints that are looked at. These endpoints include behavioral, physiological, biochemical, histological changes.

There are a number of effects that occur when an organism is simultaneously exposed to two or more toxicants. These effects include additive effects, synergistic effects, potentiation effects, and antagonistic effects. An additive effect occurs when combined effect is equal to a combination or sum of the individual effects. A synergistic effect occurs when the combination of effects is much greater than the two individual effects added together. Potentiation is an effect that occurs when an individual chemical has no effect is added to a toxicant and the combination has a greater effect than just the toxicant alone. Finally, an antagonistic effect occurs when a combination of chemicals has less of an effect than the sum of their individual effects.

Streeter–phelps Equation

Example of a stream in Arkhangelsk Oblast, Russia.

Example of a river, Tigris River near Hasankeyf, in Turkey.

The Streeter–Phelps equation is used in the study of water pollution as a water quality modelling tool. The model describes how dissolved oxygen (DO) decreases in a river or stream along a certain distance by degradation of biochemical oxygen demand (BOD). The equation was derived by H. W. Streeter, a sanitary engineer, and Earle B. Phelps, a consultant for the U.S. Public Health Service, in 1925, based on field data from the Ohio River. The equation is also known as the DO sag equation.

The Streeter–Phelps equation determines the relation between the dissolved oxygen concentration and the biological oxygen demand over time and is a solution to the linear first order differential equation

$$\frac{\partial D}{\partial t} = k_1 L_t - k_2 D$$

This differential equation states that the total change in oxygen deficit (D) is equal to the difference between the two rates of deoxygenation and reaeration at any time.

The Streeter–Phelps equation, assuming a plug-flow stream at steady state is then

$$D = \frac{k_1 L_a}{k_2 - k_1}(e^{-k_1 t} - e^{-k_2 t}) + D_a e^{-k_2 t}$$

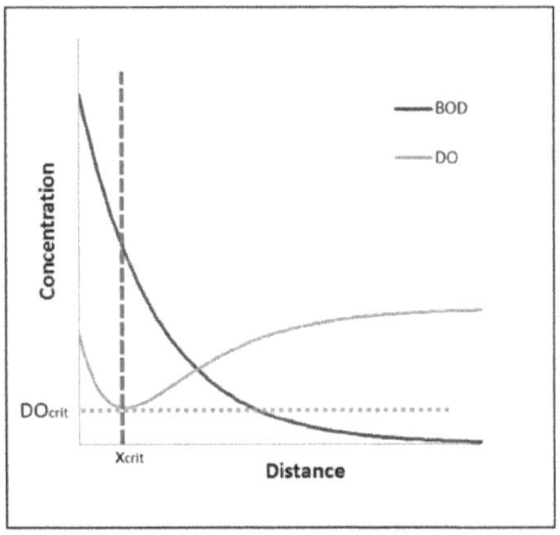

Streeter-Phelps DO sag curve and BOD development.

where

- D is the saturation deficit, which can be derived from the dissolved oxygen concentration at saturation minus the actual dissolved oxygen concentration ($D = DO_{sat} - DO$). D has the dimensions $\frac{g}{m^3}$.
- k_1 is the deoxygenation rate, usually in d^{-1}.
- k_2 is the reaeration rate, usually in d^{-1}.
- L_a is the initial oxygen demand of organic matter in the water, also called the ultimate BOD (BOD at time t=infinity). The unit of L_a is $\frac{g}{m^3}$.
- L_t is the oxygen demand remaining at time t, $L_t = L_a e^{-k_1 t}$.
- D_a is the initial oxygen deficit $[\frac{g}{m^3}]$.
- t is the elapsed time, usually $[d]$.

k_1 lies typically within the range 0.05-0.5 d^{-1} and k_2 lies typically within the range 0.4-1.5 d^{-1}. The Streeter–Phelps equation is also known as the DO sag equation. This is due to the shape of the graph of the DO over time.

Critical Oxygen Deficit

On the DO sag curve a minimum concentration occurs at some point, along a stream. If the Streeter–Phelps equation is differentiated with respect to time, and set equal to zero, the time at which the minimum DO occurs is expressed by

$$t_{crit} = \frac{1}{k_2 - k_1} \ln\left[\frac{k_2}{k_1}\left(1 - \frac{D_a(k_2 - k_1)}{L_a k_1}\right)\right]$$

To find the value of the critical oxygen deficit, D_{crit}, the Streeter–Phelps equation is combined with the equation above, for the critical time, t_{crit}. Then the minimum dissolved oxygen concentration is

$$DO_{crit} = DO_{sat} - D_{crit}$$

It should be noted that, mathematically it is possible to get a negative value of DO_{crit}, even though it is not possible to have a negative amount of DO in reality.

The distance traveled in a river from a given point source pollution or waste discharge downstream to the DO_{crit} (which is the minimum DO) is found by

$$x_{crit} = vt_{crit}$$

where v is the flow velocity of the stream. This formula is a good approximation as long as the flow can be regarded as a plug flow (turbulent).

Estimation of Reaeration Rate

Several estimations of the reaeration rate exist, which generally follow the equation

$$k_2 = Kv^a H^{-b}$$

where

- K is a constant.
- v is the flow velocity [m/s].
- H is the depth [m].
- a is a constant.
- b is a constant.

The constants depend on the system to which the equation is applied, i.e. the flow velocity and the size of the stream or river. Different values are available in the literature.

The software "International Hydrological Programme" applies the following equation derived on the basis of values used in published literature

$$k_2 = 2.148 v^{0.878} H^{-1.48}$$

where

- $k_2 \, [d^{-1}]$.
- v is the average flow velocity [m/s].
- H is the average depth of flow in the river [m].

Temperature Correction

Both the deoxygenation rate, k_1 and reaeration rate, k_2 can be temperature corrected, following the general formula.

$$k = k_{20} \theta^{(T-20)}$$

where

- k_{20} is the rate at 20 degrees Celsius.
- θ is a constant, which differs for the two rates.
- is the actual temperature in the stream in degC.

Normally θ has the value 1.048 for k_1 and 1.024 for k_2. An increasing temperature has the most impact on the deoxygenation rate, and results in an increased critical deficit (D_{crit}), and x_{crit} decreases. Furthermore, a decreased DO_{sat} concentration occurs with increasing temperature, which leads to a decrease in the DO concentration.

Mixing of Rivers

When two streams or rivers merge or water is discharged to a stream it is possible to determine the BOD and DO after mixing assuming steady state conditions and instantaneous mixing. The two streams are considered as dilutions of each other thus the initial BOD and DO will be

$$L_a = \frac{L_s Q_s + L_b Q_b}{Q_s + Q_b}$$

and

$$DO_0 = \frac{DO_s Q_s + DO_b Q_b}{Q_s + Q_b}$$

where

- L_a is the initial concentration of BOD in the river downstream of the mixing, also called BOD(0). The unit of L_a is $\frac{\text{g}}{\text{m}^3}$.
- L_b is the background BOD of the concentration in the river $[\frac{\text{g}}{\text{m}^3}]$.

- L_s is the BOD of the content of the merging river $[\frac{g}{m^3}]$.
- DO_0 is the initial concentration of the dissolved oxygen in the river downstream of the conjoining point $[\frac{g}{m^3}]$.
- DO_b is the background concentration of the dissolved oxygen content in the river $[\frac{g}{m^3}]$.
- DO_s is the background concentration of the dissolved oxygen content in the merging river $[\frac{g}{m^3}]$.
- Q_b is the flow in the river upstream from the mixing point $[\frac{m^3}{s}]$.
- Q_s is the flow in the merging river upstream from the mixing point $[\frac{m^3}{s}]$.

Numerical Approach

Nowadays it is possible to solve the classical Streeter–Phelps equation numerically by use of computers. The differential equations are solved by integration.

History

In 1925, a study on the phenomena of oxidation and reaeration in the Ohio River in the US was published by the sanitary engineer, Harold Warner Streeter and the consultant, Earle Bernard Phelps (1876–1953). The study was based on data obtained from May 1914 to April 1915 by the United States Public Health Service under supervision of Surg. W.H. Frost.

More complex versions of the Streeter-Phelps model were introduced during the 1960s, where computers made it possible to include further contributions to the oxygen development in streams. At the head of this development were O'Connor (1960) and Thomann (1963). O'Connor added the contributions from photosynthesis, respiration and sediment oxygen demand (SOD). Thomann expanded the Streeter-Phelps model to allow for multi segment systems.

Applications and Limitations

The simple Streeter-Phelps model is based on the assumptions that a single BOD input is distributed evenly at the cross section of a stream or river and that it moves as plug flow with no mixing in the river. Furthermore only one DO sink (carbonaceous BOD) and one DO source (reaeration) is considered in the classical Streeter-Phelps model. These simplifications will give rise to errors in the model. For example the model does not include BOD removal by sedimentation, that suspended BOD is converted to a dissolved state, that sediment has an oxygen demand and that photosynthesis and respiration will impact the oxygen balance.

Expanded Model

In addition to the oxidation of organic matter and the reaeration process, there are many other processes in a stream which affect the DO. In order to make a more accurate model it is possible to include these factors using an expanded model.

The expanded model is a modification of the traditional model and includes internal sources (re-aeration and photosynthesis) and sinks (BOD, background BOD, SOD and respiration) of DO. It is not always necessary to include all of these parameters. Instead relevant sources and sinks can be summed to yield the overall solution for the particular model. Parameters in the expanded model can be either measured in the field or estimated theoretically.

Background BOD

Background BOD or benthic oxygen demand is the diffuse source of BOD represented by the decay of organic matter that has already settled on the bottom. This will give rise to a constant diffuse input thus the change in BOD over time will be

$$\frac{dL}{dt} = -k_1 L + L_b$$

where

- k_1 is the rate for oxygen consumption by BOD, usually in d^{-1}.
- L is the BOD from organic matter in the water [—].
- L_b is the background BOD input $[\frac{g}{m^3}]$.

Sedimentation of BOD

Sedimented BOD does not directly consume oxygen and this should therefore be taken into account. This is done by introducing a rate of BOD removal combined with a rate of oxygen consumption by BOD. Giving a total rate for oxygen removal by BOD

$$k_r = k_1 + k_3$$

where

- k_1 is the rate of oxygen consumption by BOD, usually in d^{-1}.
- k_3 is the rate of settling of BOD, usually in d^{-1}.

The change in BOD over time is described as

$$\frac{dL}{dt} = -k_r L$$

where L is the BOD from organic matter in the water $[\frac{g}{m^3}]$.

k_r is typically in the range of 0.5-5 d^{-1}.

Sediment Oxygen Demand

Oxygen can be consumed by organisms in the sediment. This process is referred to as sediment oxygen demand (SOD). Measurement of SOD can be undertaken by measuring the change of oxygen in a box on the sediment.(benthic respirometer)

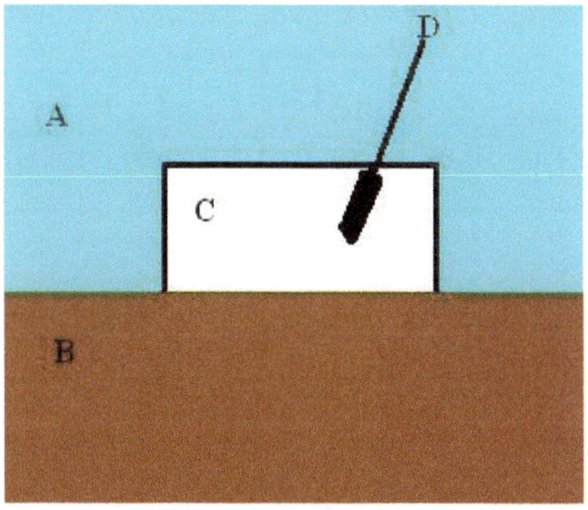

Measurement of SOD using an oxygen electrode. A: Water, B: Sediment, C: Box, D: Oxygen electrode.

The change in oxygen deficit due to consumption by sediment is described as

$$\frac{dD}{dt} = -k_2 D + \frac{S}{H}$$

where

- H is the depth of the river [m]
- S is the SOD [$\frac{g}{m^2 d}$]
- D is the saturation deficit [$\frac{g}{m^3}$].
- k_2 is the reaeration rate [d^{-1}].

The range of the SOD is typically in the range of 0.1 – 1 $\frac{g}{m^2 d}$ for a natural river with low pollution and 5 – 10 $\frac{g}{m^2 d}$ for a river with moderate to heavy pollution.

Nitrification

Ammonium is oxidized to nitrate under aerobic conditions

$$NH_4^+ + 2O_2 \rightarrow NO_3^- + H_2O + 2H^+$$

Ammonium oxidation can be treated as part of BOD, so that BOD = CBOD + NBOD, where CBOD is the carbonaceous biochemical oxygen demand and NBOD is nitrogenous BOD. Usually CBOD is much higher than the ammonium concentration and thus NBOD often does not need to be considered. The change in oxygen deficit due to oxidation of ammonium is described as

$$\frac{dD}{dt} = k_N N - k_2 D$$

where

- D is the saturation deficit.
- k_N is the nitrification rate [d^{-1}].
- N is ammonium-nitrogen concentration.

The range of k_N is typically 0.05-0.5 d^{-1}.

Photosynthesis and Respiration

Photosynthesis and respiration are performed by algae and by macrophytes. Respiration is also performed by bacteria and animals. Assuming steady state (net daily average) the change in deficit will be

$$\frac{dD}{dt} = -k_2 D + (R+P)_{avg}$$

where

- R is the respiration $[\frac{mg}{Ld}]$.
- P is the photosynthesis $[\frac{mg}{Ld}]$.

Note that BOD only includes respiration of microorganisms e.g. algae and bacteria and not by macrophytes and animals.

Due to the variation of light over time, the variation of the photosynthetic oxygen can be described by a periodical function over time, where time is after sunrise and before sunset

$$P(t) = P_{max} sin\left(\frac{\pi}{f} + (t - t_s)\right)$$

where

- $P(t)a$ is the photosynthesis at a given time $[\frac{mg}{Ld}]$.
- P_{max} is the daily maximum of the photosynthesis $[\frac{mg}{Ld}]$.
- f is the fraction of day with sunlight, usually $\frac{1}{2}$ day.
- t_s is the time at which sun rises $[d]$.

The range of the daily average value of primary production $(P-R)$ is typically 0.5-10 $\frac{mg}{Ld}$.

This contribution was made originally as part of the Water Pollution course at the Technical University of Denmark in 2011.

Sources of Water Pollution

Types, Effects and Sources of Water Pollution

- Point sources
- Nonpoint sources
- Water quality

Nonpoint Source Pollution

Nonpoint source (NPS) pollution is a term used to describe pollution resulting from many diffuse sources, in direct contrast to point source pollution which results from a single source. Nonpoint source pollution generally results from land runoff, precipitation, atmospheric deposition, drainage, seepage, or hydrological modification (rainfall or snowmelt) where tracing the pollution back to a single source is difficult.

Muddy river

Non-point source water pollution affects a water body from sources such as polluted runoff from agricultural areas draining into a river, or wind-borne debris blowing out to sea. Non-point source air pollution affects air quality from sources such as smokestacks or car tailpipes. Although these pollutants have originated from a point source, the long-range transport ability and multiple sources of the pollutant make it a non-point source of pollution. Non-point source pollution can be contrasted with point source pollution, where discharges occur to a body of water or into the atmosphere at a single location.

NPS may derive from many different sources with no specific solution may change to rectify the problem, making it difficult to regulate. Non point source water pollution is difficult to control because it comes from the everyday activities of many different people, such as fertilizing a lawn, using a pesticide, or constructing a road or building.

It is the leading cause of water pollution in the United States today, with polluted runoff from agriculture the primary cause.

Other significant sources of runoff include hydrological and habitat modification, and silviculture (forestry).

Contaminated stormwater washed off parking lots, roads and highways, and lawns (often containing fertilizers and pesticides) is called urban runoff. This runoff is often classified as a type of NPS pollution. Some people may also consider it a point source because many times it is channeled into municipal storm drain systems and discharged through pipes to nearby surface waters. However, not all urban runoff flows through storm drain systems before entering water bodies. Some may flow directly into water bodies, especially in developing and suburban areas. Also, unlike other types of point sources, such as industrial discharges, sewage treatment plants and other operations, pollution in urban runoff cannot be attributed to one activity or even group of activities. Therefore, because it is not caused by an easily identified and regulated activity, urban runoff pollution sources are also often treated as true non-point sources as municipalities work to abate them.

Principal Types

Runoff of soil and fertilizer during a rain storm

Sediment

Sediment (loose soil) includes silt (fine particles) and suspended solids (larger particles). Sediment may enter surface waters from eroding stream banks, and from surface runoff due to improper plant cover on urban and rural land. Sediment creates turbidity (cloudiness) in water bodies, reducing the amount of light reaching lower depths, which can inhibit growth of submerged aquatic plants and consequently affect species which are dependent on them, such as fish and shellfish. High turbidity levels also inhibit drinking water purification systems.

Sediment can also be discharged from multiple different sources. Sources include construction sites (although these are point sources, which can be managed with erosion controls and sediment controls), agricultural fields, stream banks, and highly disturbed areas.

Nutrients

Nutrients mainly refers to inorganic matter from runoff, landfills, livestock operations and crop lands. The two primary nutrients of concern are phosphorus and nitrogen.

Phosphorus is a nutrient that occurs in many forms that are bioavailable. It is notoriously over-abundant in human sewage sludge. It is a main ingredient in many fertilizers used for agriculture as well as on residential and commercial properties, and may become a limiting nutrient in freshwater systems and some estuaries. Phosphorus is most often transported to water bodies via soil erosion because many forms of phosphorus tend to be adsorbed on to soil particles. Excess amounts of phosphorus in aquatic systems (particularly freshwater lakes, reservoirs, and ponds) leads to proliferation of microscopic algae called phytoplankton. The increase of organic matter supply due to the excessive growth of the phytoplankton is called eutrophication. A common symptom of eutrophication is algae blooms that can produce unsightly surface scums, shade out beneficial types of plants, and poison the water due to toxins produced by the algae. These toxins are a particular problem in systems used for drinking water because some toxins can cause human illness and removal of the toxins is difficult and expensive. Bacterial decomposition of algal blooms consumes dissolved oxygen in the water, generating hypoxia with detrimental consequences for fish and aquatic invertebrates.

Nitrogen is the other key ingredient in fertilizers, and it generally becomes a pollutant in saltwater or brackish estuarine systems where nitrogen is a limiting nutrient. Similar to phosphorus in fresh-waters, excess amounts of bioavailable nitrogen in marine systems lead to eutrophication and algae blooms. Hypoxia is an increasingly common result of eutrophication in marine systems and can impact large areas of estuaries, bays, and near shore coastal waters. Each summer, hypoxic conditions form in bottom waters where the Mississippi River enters the Gulf of Mexico. During recent summers, the aereal extent of this "dead zone" is comparable to the area of New Jersey and has major detrimental consequences for fisheries in the region.

Nitrogen is most often transported by water as nitrate (NO_3). The nitrogen is usually added to a watershed as organic-N or ammonia (NH_3), so nitrogen stays attached to the soil until oxidation converts it into nitrate. Since the nitrate is generally already incorporated into the soil, the water traveling through the soil (i.e., interflow and tile drainage) is the most likely to transport it, rather than surface runoff.

Toxic Contaminants and Chemicals

Compounds including heavy metals like lead, mercury, zinc, and cadmium, organics like polychlorinated biphenyls (PCBs) and polycyclic aromatic hydrocarbons (PAHs), fire retardants, and other substances are resistant to breakdown. These contaminants can come from a variety of sources including human sewage sludge, mining operations, vehicle emissions, fossil fuel combustion, urban runoff, industrial operations and landfills.

Toxic chemicals mainly include organic compounds and inorganic compounds. These compounds include pesticides like DDT, acids, and salts that have severe effects to the ecosystem and water-bodies. These compounds can threaten the health of both humans and aquatic species while being resistant to environmental breakdown, thus allowing them to persist in the environment. These toxic chemicals could come from croplands, nurseries, orchards, building sites, gardens, lawns and landfills.

Acids and salts mainly are inorganic pollutants from irrigated lands, mining operations, urban runoff, industrial sites and landfills.

Pathogens

Pathogens are bacteria and viruses that can be found in water and cause diseases in humans. Typically, pathogens cause disease when they are present in public drinking water supplies. Pathogens found in contaminated runoff may include:

- *Cryptosporidium parvum*
- *Giardia lamblia*
- *Salmonella*
- *Novovirus* and other viruses
- Parasitic worms (helminths).

Coliform bacteria and fecal matter may also be detected in runoff. These bacteria are a commonly used indicator of water pollution, but not an actual cause of disease.

Pathogens may contaminate runoff due to poorly managed livestock operations, faulty septic systems, improper handling of pet waste, the over application of human sewage sludge, contaminated storm sewers, and sanitary sewer overflows.

Principal Sources

Urban and Suburban Areas

Urban and suburban areas are a main sources of nonpoint source pollution due to the amount of runoff that is produced due to the large amount of paved surfaces. Paved surfaces, such as asphalt and concrete are impervious to water penetrating them. Any water that is on contact with these surfaces will run off and be absorbed by the surrounding environment. These surfaces make it easier for stormwater to carry pollutants into the surrounding soil.

Construction sites tend to have disturbed soil that is easily eroded by precipitation like rain, snow, and hail. Additionally, discarded debris on the site can be carried away by runoff waters and enter the aquatic environment.

Typically, in suburban areas, chemicals are used for lawn care. These chemicals can end up in runoff and enter the surrounding environment via storm drains in the city. Since the water in storm drains is not treated before flowing into surrounding water bodies, the chemicals enter the water directly.

Agricultural Operations

Agricultural operations account for a large percentage of all nonpoint source pollution in the United States. When large tracts of land are plowed to grow crops, it exposes and loosens soil that was once buried. This makes the exposed soil more vulnerable to erosion during rainstorms. It also can increase the amount of fertilizer and pesticides carried into nearby bodies of water.

Atmospheric Inputs

Atmospheric inputs of pollution into the air can come from multiple sources. Typically, industrial facilities, like factories, emit air pollution via a smokestack. Although this is a point source, due to the distributional nature, long-range transport, and multiple sources of the pollution, it is considered a nonpoint source. Additionally, atmospheric pollution can become water pollution, by being washed out of the atmosphere in the form of rain or snow.

Highway Runoff

Highway runoff accounts for a small but widespread percentage of all nonpoint source pollution. Harned (1988) estimated that runoff loads were composed of atmospheric fallout (9%), vehicle deposition (25%) and highway maintenance materials (67%) he also estimated that about 9 percent of these loads were reentrained in the atmosphere.

Forestry and Mining Operations

Forestry and mining operations can have significant inputs to non-point source pollution.

Forestry

Forestry operations reduce the number of trees in a given area, thus reducing the oxygen levels in that area as well. This action, coupled with the heavy machinery rolling over the soil increases the risk of erosion.

Mining

Active mining operations are considered point sources, however runoff from abandoned mining operations contribute to nonpoint source pollution. In strip mining operations, the top of the mountain is removed to expose the desired ore. If this area is not properly reclaimed once the mining has finished, soil erosion can occur. Additionally, there can be chemical reactions with the air and newly exposed rock to create acidic runoff. Water that seeps out of abandoned subsurface mines can also be highly acidic. This can seep into the nearest body of water and change the pH in the aquatic environment.

Marinas and Boating Activities

Chemicals used for boat maintenance, like paint, solvents, and oils find their way into water through runoff. Additionally, spilling fuels or leaking fuels directly into the water from boats contribute to nonpoint source pollution. Nutrient and bacteria levels are increased by poorly maintained sanitary waste receptacles on the boat and pump-out stations.

Control

Contour buffer strips used to retain soil and reduce erosion

Urban and Suburban Areas

To control non-point source pollution, many different approaches can be undertaken in both urban and suburban areas. Buffer strips provide a barrier of grass in between impervious paving material like parking lots and roads, and the closest body of water. This allows the soil to absorb any pollution before it enters the local aquatic system. Retention ponds can be built in drainage areas to create an aquatic buffer between runoff pollution and the aquatic environment. Runoff and storm water drain into the retention pond allowing for the contaminants to settle out and become trapped in the pond.

The use of porous pavement allows for rain and storm water to drain into the ground beneath the pavement, reducing the amount of runoff that drains directly into the water body. Restoration methods such as constructing wetlands are also used to slow runoff as well as absorb contamination.

Construction sites typically implement simple measures to reduce pollution and runoff. Firstly, sediment or silt fences are erected around construction sites to reduce the amount of sediment and large material draining into the nearby water body. Secondly, laying grass or straw along the border of construction sites also work to reduce nonpoint source pollution.

In areas served by single-home septic systems, local government regulations can force septic system maintenance to ensure compliance with water quality standards. In Washington (state), a novel approach was developed through a creation of a "shellfish protection district" when either a commercial or recreational shellfish bed is downgraded because of ongoing nonpoint source pollution. The shellfish protection district is a geographic area designated by a county to protect water quality and tideland resources, and provides a mechanism to generate local funds for water quality services to control nonpoint sources of pollution. At least two shellfish protection districts in south Puget Sound have instituted septic system operation and maintenance requirements with program fees tied directly to property taxes.

Agricultural Operations

To control sediment and runoff, farmers may utilize erosion controls to reduce runoff flows and retain soil on their fields. Common techniques include contour plowing, crop mulching, crop rotation, planting perennial crops and installing riparian buffers. *Conservation tillage* is a concept used to reduce runoff while planting a new crop. The farmer leaves some crop reside from the previous planting in the ground to help prevent runoff during the planting process.

Nutrients are typically applied to farmland as commercial fertilizer; animal manure; or spraying of municipal or industrial wastewater (effluent) or sludge. Nutrients may also enter runoff from crop residues, irrigation water, wildlife, and atmospheric deposition. Farmers can develop and implement nutrient management plans to reduce excess application of nutrients.

To minimize pesticide impacts, farmers may use Integrated Pest Management (IPM) techniques (which can include biological pest control) to maintain control over pests, reduce reliance on chemical pesticides, and protect water quality.

Forestry Operations

With a well-planned placement of both logging trails, also called skid trails, can reduce the amount of sediment generated. By planning the trails location as far away from the logging activity as possible as well as contouring the trails with the land, it can reduce the amount of loose sediment in the runoff. Additionally, by replanting trees on the land after logging, it provides a structure for the soil to regain stability as well as replaces the logged environment.

Marinas

Installing shut off valves on fuel pumps at a marina dock can help reduce the amount of spillover into the water. Additionally, pump-out stations that are easily accessible to boaters in a marina can provide a clean place in which to dispose of sanitary waste without dumping it directly into the water. Finally, something as simple as having trash containers around a marina can prevent larger objects entering the water.

Groundwater Pollution

Groundwater pollution example in Lusaka, Zambia where the pit latrine in the background is polluting the shallow well in the foreground with pathogens and nitrate.

Groundwater pollution (also called groundwater contamination) occurs when pollutants are released to the ground and make their way down into groundwater. It can also occur naturally due to the presence of a minor and unwanted constituent, contaminant or impurity in the groundwater, in which case it is more likely referred to as contamination rather than pollution.

The pollutant creates a contaminant plume within an aquifer. Movement of water and dispersion within the aquifer spreads the pollutant over a wider area. Its advancing boundary, often called a plume edge, can intersect with groundwater wells or daylight into surface water such as seeps and spring, making the water supplies unsafe for humans and wildlife. The movement of the plume, called a plume front, may be analyzed through a hydrological transport model or groundwater model. Analysis of groundwater pollution may focus on soil characteristics and site geology, hydrogeology, hydrology, and the nature of the contaminants.

Pollution can occur from on-site sanitation systems, landfills, effluent from wastewater treatment plants, leaking sewers, petrol filling stations or from over application of fertilizers in agriculture. Pollution (or contamination) can also occur from naturally occurring contaminants, such as arsenic or fluoride. Using polluted groundwater causes hazards to public health through poisoning or the spread of disease.

Different mechanisms have influence on the transport of pollutants, e.g. diffusion, adsorption, precipitation, decay, in the groundwater. The interaction of groundwater contamination with surface waters is analyzed by use of hydrology transport models.

Pollutant Types

Contaminants found in groundwater cover a broad range of physical, inorganic chemical, organic chemical, bacteriological, and radioactive parameters. Principally, many of the same pollutants that play a role in surface water pollution may also be found in polluted groundwater, although their respective importance may differ.

Arsenic and Fluoride

Arsenic and fluoride have been recognized by the World Health Organization (WHO) as the most serious inorganic contaminants in drinking-water on a worldwide basis.

The metalloid arsenic can occur naturally in groundwater, as seen most frequently in Asia, including in China, India and Bangladesh. In the Ganges Plain of northern India and Bangladesh severe contamination of groundwater by naturally occurring arsenic affects 25% of water wells in the shallower of two regional aquifers.

Arsenic in groundwater can also be present where there are mining operations or mine waste dumps that will leach arsenic.

Natural fluoride in groundwater is of growing concern as deeper groundwater is being used, "with more than 200 million people at risk of drinking water with elevated concentrations." Fluoride can especially be released from acidic volcanic rocks and dispersed volcanic ash when water hardness is low. High levels of fluoride in groundwater is a serious problem in the Argentinean Pampas, Chile, Mexico, India, Pakistan, the East African Rift, and some volcanic islands (Tenerife)

In areas that have naturally occurring high levels of fluoride in groundwater which is used for drinking water, both dental and skeletal fluorosis can be prevalent and severe.

Pathogens

Waterborne diseases can be spread via a groundwater well which is contaminated with fecal pathogens from pit latrines

Pathogens contained in feces can lead to groundwater pollution when they are given the opportunity to reach the groundwater, making it unsafe for drinking. Of the four pathogen types that are present in feces (bacteria, viruses, protozoa and helminths or helminth eggs), the first three can be commonly found in polluted groundwater, whereas the relatively large helminth eggs are usually filtered out by the soil matrix.

Groundwater that is contaminated with pathogens can lead to fatal fecal-oral transmission of diseases (e.g. cholera, diarrhoea).

Nitrate

Nitrate is the most common chemical contaminant in the world's groundwater and aquifers. In some low-income countries nitrate levels in groundwater are extremely high, causing significant health problems. It is also stable (it does not degrade) under high oxygen conditions.

Nitrate levels above 10 mg/L (10 ppm) in groundwater can cause "blue baby syndrome" (acquired methemoglobinemia). Drinking water quality standards in the European Union stipulate less than 50 mg/L for nitrate in drinking water.

However, the linkages between nitrates in drinking water and blue baby syndrome have been disputed in other studies. The syndrome outbreaks might be due to other factors than elevated nitrate concentrations in drinking water.

Elevated nitrate levels in groundwater can be caused by on-site sanitation, sewage sludge disposal and agricultural activities. It can therefore have an urban or agricultural origin.

Organic Compounds

Volatile organic compounds (VOCs) are a dangerous contaminant of groundwater. They are generally introduced to the environment through careless industrial practices. Many of these compounds were not known to be harmful until the late 1960s and it was some time before regular testing of groundwater identified these substances in drinking water sources.

Primary VOC pollutants found in groundwater include aromatic hydrocarbons such as BTEX compounds (benzene, toluene, ethylbenzene and xylenes), and chlorinated solvents including tetrachloroethylene (PCE), trichloroethylene (TCE), and vinyl chloride (VC). BTEX are important components of gasoline. PCE and TCE are industrial solvents historically used in dry cleaning processes and as a metal degreaser, respectively.

Other organic pollutants present in groundwater and derived from industrial operations are the polycyclic aromatic hydrocarbons (PAHs). Due to its molecular weight, Naphthalene is the most soluble and mobile PAH found in groundwater, whereas benzo(a)pyrene is the most toxic one. PAHs are generally produced as byproducts by incomplete combustion of organic matter.

Organic pollutants can also be found in groundwater as insecticides and herbicides. As many other synthetic organic compounds, most pesticides have very complex molecular structures. This complexity determines the water solubility, adsorption capacity, and mobility of pesticides in the groundwater system. Thus, some types of pesticides are more mobile than others so they can more easily reach a drinking-water source.

Metals

Several trace metals can occurs naturally in some rocks or enter in the environmental from natural processes such as weathering. However, industrial activities such as mining and metallurgy, solid waste disposal, painting and enamel works can lead to severe groundwater pollution with elevated concentrations of toxic metals including lead, cadmium and chromium.

The migration of metals (and metalloids) in groundwater will be affected by several factors, in particular by chemical reactions which determine the partitioning of contaminants among different phases and species. Thus, the mobility of metals primarily depends on the pH and redox state of groundwater.

Others

Other organic pollutants include a range of organohalides and other chemical compounds, petroleum hydrocarbons, various chemical compounds found in personal hygiene and cosmetic products, drug pollution involving pharmaceutical drugs and their metabolites. Inorganic pollutants might include other nutrients such as ammonia and phosphate, and radionuclides such as uranium (U) or radon (Rn) naturally present in some geological formations. Saltwater intrusion is also an example of natural contamination, but is very often intensified by human activities.

Groundwater pollution is a worldwide issue. A study of the groundwater quality of the principal aquifers of the United States conducted between 1991 and 2004, showed that 23% of domestic wells had contaminants at levels greater than human-health benchmarks. Another study suggested that the major groundwater pollution problems in Africa, considering the order of importance are: (1) nitrate pollution, (2) pathogenic agents, (3) organic pollution, (4) salinization, and (5) acid mine drainage.

Causes

Naturally-occurring (Geogenic)

"Geogenic" refers to naturally occurring as a result from geological processes.

The natural arsenic pollution occurs because aquifer sediments contain organic matter that generates anaerobic conditions in the aquifer. These conditions result in the microbial dissolution of iron oxides in the sediment and, thus, the release of the arsenic, normally strongly bound to iron oxides, into the water. As a consequence, arsenic-rich groundwater is often iron-rich, although secondary processes often obscure the association of dissolved arsenic and dissolved iron. Arsenic is found in groundwater most commonly as the reduced species arsenite and the oxidized species arsenate, being the acute toxicity of arsenite somewhat greater than that of arsenate. Investigations by WHO indicated that 20% of 25,000 boreholes tested in Bangladesh had arsenic concentrations exceeding 50 µg/l.

The occurrence of fluoride is close related to the abundance and solubility of fluoride-containing minerals such as fluorite ($CaF2$). Considerably high concentrations of fluoride in groundwater are typically caused by a lack of calcium in the aquifer. Health problems associated with dental fluorosis may occur when fluoride concentrations in groundwater exceed 1.5 mg/l, which is the WHO guideline value since 1984.

The Swiss Federal Institute of Aquatic Science and Technology (EAWAG) has recently developed the interactive Groundwater Assessment Platform (GAP), where the geogenic risk of contamination in a given area can be estimated using geological, topographical and other environmental data without having to test samples from every single groundwater resource. This tool also allows the user to produce probability risk mapping for both arsenic and fluoride.

High concentrations of parameters like salinity, iron, manganese, uranium, radon and chromium, in groundwater, may also be of geogenic origin. This contaminants can be important locally but they are not as widespread as arsenic and fluoride.

On-site Sanitation Systems

A traditional housing compound near Herat, Afghanistan, where a shallow water supply well (foreground) is in close proximity to the pit latrine (behind the white greenhouse) leading to contamination of the groundwater.

Groundwater pollution with pathogens and nitrate can also occur from the liquids infiltrating into the ground from on-site sanitation systems such as pit latrines and septic tanks, depending on the population density and the hydrogeological conditions.

Factors controlling the fate and transport of pathogens are quite complex and the interaction among them is not well understood. If the local hydrogeological conditions (which can vary within a space of a few square kilometres) are ignored, simple on-site sanitation infrastructures such as pit latrines can cause significant public health risks via contaminated groundwater.

Liquids leach from the pit and pass the unsaturated soil zone (which is not completely filled with water). Subsequently, these liquids from the pit enter the groundwater where they may lead to groundwater pollution. This is a problem if a nearby water well is used to supply groundwater for drinking water purposes. During the passage in the soil, pathogens can die off or be adsorbed significantly, mostly depending on the travel time between the pit and the well. Most, but not all pathogens die within 50 days of travel through the subsurface.

The degree of pathogen removal strongly varies with soil type, aquifer type, distance and other environmental factors. For example, the unsaturated zone becomes "washed" during extended periods of heavy rain, providing hydraulic pathway for the quick pass of pathogens. It is difficult to estimate the safe distance between a pit latrine or a septic tank and a water source. In any case, such recommendations about the safe distance are mostly ignored by those building pit latrines. In addition, household plots are of a limited size and therefore pit latrines are often built much closer to groundwater wells than what can be regarded as safe. This results in groundwater pollution and household members falling sick when using this groundwater as a source of drinking water.

Sewage (Treated and Untreated)

Groundwater pollution can be caused by untreated waste discharge leading to diseases like skin lesions, bloody diarrhea and dermatitis. This is more common in locations having limited wastewater treatment infrastructure, or where there are systematic failures of the on-site sewage disposal system. Along with pathogens and nutrients, untreated sewage can also have an important load of heavy metals that may seep into the groundwater system.

The treated effluent from sewage treatment plants may also reach the aquifer if the effluent is infiltrated or discharged to local surface water bodies. Therefore, those substances that are not removed in conventional sewage treatment plants may reach the groundwater as well. For example, detected concentrations of pharmaceutical residues in groundwater were in the order of 50 ng/L in several locations in Germany. This is because in conventional sewage treatment plants, micro-pollutants such as hormones, pharmaceutical residues and other micro-pollutants contained in urine and feces are only partially removed and the remainder is discharged into surface water, from where it may also reach the groundwater.

Groundwater pollution can also occur from leaking sewers which has been observed for example in Germany. This can also lead to potential cross-contamination of drinking-water supplies.

Spreading wastewater or sewage sludge in agriculture may also be included as sources of faecal contamination in groundwater.

Fertilizers and Pesticides

Nitrate can also enter the groundwater via excessive use of fertilizers, including manure spreading. This is because only a fraction of the nitrogen-based fertilizers is converted to produce and other plant matter. The remainder accumulates in the soil or lost as run-off. High application rates of nitrogen-containing fertilizers combined with the high water-solubility of nitrate leads to increased runoff into surface water as well as leaching into groundwater, thereby causing groundwater pollution. The excessive use of nitrogen-containing fertilizers (be they synthetic or natural) is particularly damaging, as much of the nitrogen that is not taken up by plants is transformed into nitrate which is easily leached.

The nutrients, especially nitrates, in fertilizers can cause problems for natural habitats and for human health if they are washed off soil into watercourses or leached through soil into groundwater. The heavy use of nitrogenous fertilizers in cropping systems is the largest contributor to anthropogenic nitrogen in groundwater worldwide.

Feedlots/animal corrals can also lead to the potential leaching of nitrogen and metals to groundwater. Over application of animal manure may also result in groundwater pollution with pharmaceutical residues derived from veterinary drugs.

The US Environmental Protection Agency (EPA) and the European Commission are seriously dealing with the nitrate problem related to agricultural development, as a major water supply problem that requires appropriate management and governance.

Runoff of pesticides may leach into groundwater causing human health problems from contaminated water wells. Pesticide concentrations found in groundwater are typically low, and often the regulatory human health-based limits exceeded are also very low. The organophosphorus insecticide monocrotophos (MCP) appears to be one of a few hazardous, persistent, soluble and mobile (it does not bind with minerals in soils) pesticides able to reach a drinking-water source. In general, more pesticide compounds are being detected as groundwater quality monitoring programs have become more extensive; however, much less monitoring has been conducted in developing countries due to the high analysis costs.

Commercial and Industrial Leaks

A wide variety of both inorganic and organic pollutants have been found in aquifers underlying commercial and industrial activities.

Ore mining and metal processing facilities are the primary responsible of the presence of metals in groundwater of anthropogenic origin, including arsenic. The low pH associated with acid mine drainage (AMD) contributes to the solubility of potential toxic metals that can eventually enter the groundwater system.

There is a increasing concern over the groundwater pollution by gasoline leaked from petroleum underground storage tanks (USTs) of gas stations. BTEX compounds are the most common additives of the gasoline. BTEX compounds, including benzene, have densities lower than water (1 g/ml). Similar to the oil spills on the sea, the non-miscible phase, referred to as Light Non-Aqueous Phase Liquid (LNAPL), will "float" upon the water table in the aquifer.

Chlorinated solvents are used in nearly any industrial practice where degreasing removers are required. PCE is a highly utilized solvent in the dry cleaning industry because of its cleaning effectiveness and relatively low cost. It has also been used for metal-degreasing operations. Because it is is highly volatile, it is more frequently found in groundwater than in surface water. TCE has historically been used as a metal cleaning. The military facility Anniston Army Dept (ANAD) in the United States was placed on the US EPA Superfund National Priorities List (NPL) because of groundwater contamination with as much as 27 million pounds of TCE. Both PCE and TCE may degrade to vinyl chloride (VC), the most toxic chlorinated hydrocarbon.

Many types of solvents may have also been disposed illegally, leaking over time to the groundwater system.

Chlorinated solvents such as PCE and TCE have densities higher than water and the non-miscible phase is referred to as Dense Non-Aqueous Phase Liquids (DNAPL). Once they reach the aquifer, they will "sink" and eventually accumulate on the top of low-permeability layers.

Historically, wood-treating facilities have also release insecticides such as pentachlorophenol (PCP) and creosote into the environment, impacting the groundwater resources. PCP is a a highly soluble and toxic obsolete pesticide recently listed in the Stockholm Convention on Persistent Organic Pollutants. PAHs and other semi-VOCs are the common contaminants associated with creosote.

Although non-miscible, both LNAPLs and DNAPLs still have the potential to slowly dissolve into the aqueous (miscible) phase to create a plume and thus become a long-term source of contamination. DNAPLs (chlorinated solvents, heavy PAHs, creosote, PCBs) tend to be difficult to manage as they can reside very deep in the groundwater system.

Hydraulic Fracturing

The recent growth of Hydraulic Fracturing ("Fracking") wells in the United States has raised concerns regarding its potential risks of contaminating groundwater resources. The Environmental

Protection Agency (EPA), along with many other researchers, has been delegated to study the relationship between hydraulic fracturing and drinking water resources. While it is possible to perform hydraulic fracturing without having a relevant impact on groundwater resources if stringent controls und quality management measures are in place, there are a number of cases where groundwater pollution due to improper handling or technical failures was observed.

While the EPA has not found significant evidence of a widespread, systematic impact on drinking water by hydraulic fracturing, this may be due to insufficient systematic pre- and post- hydraulic fracturing data on drinking water quality, and the presence of other agents of contamination that preclude the link between shale oil/gas extraction and its impact.

Despite the EPA's lack of profound widespread evidence, other researchers have made significant observations of rising groundwater contamination in close proximity to major shale oil/gas drilling sites located in Marcellus *Ellsworth, William (2013). "Injection-Induced Earthquakes". Science* (British Columbia, Canada). Within one kilometer of these specific sites, a subset of shallow drinking water consistently showed higher concentration levels of methane, ethane, and propane concentrations than normal. An evaluation of higher Helium and other noble gas concentration along with the rise of hydrocarbon levels supports the distinction between hydraulic fracturing fugitive gas and naturally occurring "background" hydrocarbon content. This contamination is speculated to be the result of leaky, failing, or improperly installed gas well casings.

Furthermore, it is theorized that contamination could also result from the capillary migration of deep residual hyper-saline water and hydraulic fracturing fluid, slowly flowing through faults and fractures until finally making contact with groundwater resources; however, many researchers argue that the permeability of rocks overlying shale formations are too low to allow this to ever happen sufficiently. To ultimately prove this theory, there would have to be traces of toxic trihalomethanes (THM) since they are often associated with the presence of stray gas contamination, and typically co-occur with high halogen concentrations in hyper-saline waters. Besides, highly saline waters are a common natural feature in deep groundwater systems.

While conclusions regarding groundwater pollution as the result to hydraulic fracturing fluid flow is restricted in both space and time, researchers have hypothesized that the potential for systematic stray gas contamination depends mainly on the integrity of the shale oil/gas well structure, along with its relative geological location to local fracture systems that could potentially provide flow paths for fugitive gas migration.

Though widespread, systematic contamination by hydraulic fracturing has been heavily disputed, one major source of contamination that has the most consensus among researchers of being the most problematic is site-specific accidental spillage of hydraulic fracturing fluid and produced water. So far, a significant majority of groundwater contamination events are derived from surface-level anthropogenic routes rather than the subsurface flow from underlying shale formations. Examples of such events include: a fracking fluid spillage in Acorn Fork Creek, Kentucky that caused a widespread death among aquatic species in 2007; a 420,000 gallon spillage of hyper-saline produced water that turned a once very-fertile farmland in New Mexico into a dead-zone in 2010; and a 42,000 gallon fracking fluid spillage in Arlington, Texas that necessitated an evacuation of over a 100 homes in 2015. While the damage can be obvious, and much more effort is being done to prevent these accidents from occurring so frequently, the lack of data from fracking oil spills continue to leave researchers in the dark. In many of these events, the data acquired from the leakage or spillage is often very vague, and thus would lead researchers to lacking conclusions.

Researchers from the Federal Institute for Geosciences and Natural Resources (BGR) conducted a modelling study for a deep shale-gas formation in the North German Basin. They concluded that the probability is small that the rise of fracking fluids through the geological underground to the surface will impact shallow groundwater.

Landfill Leachate

Leachate from sanitary landfills can lead to groundwater pollution.

Love Canal was one of the most widely known examples of groundwater pollution. In 1978, residents of the Love Canal neighborhood in upstate New York noticed high rates of cancer and an alarming number of birth defects. This was eventually traced to organic solvents and dioxins from an industrial landfill that the neighborhood had been built over and around, which had then infiltrated into the water supply and evaporated in basements to further contaminate the air. Eight hundred families were reimbursed for their homes and moved, after extensive legal battles and media coverage.

Other

Further causes of groundwater pollution are chemical spills from commercial or industrial operations, chemical spills occurring during transport (e.g. spillage of diesel fuels), illegal waste dumping, infiltration from urban runoff or mining operations, road salts, de-icing chemicals from airports and even atmospheric contaminants since groundwater is part of the hydrologic cycle.

The burial of corpses and their subsequent degradation may also pose a risk of pollution to groundwater.

Mechanisms

The passage of water through the subsurface can provide a reliable natural barrier to contamination but it only works under favorable conditions.

The stratigraphy of the area plays an important role in the transport of pollutants. An area can have layers of sandy soil, fractured bedrock, clay, or hardpan. Areas of karst topography on limestone bedrock are sometimes vulnerable to surface pollution from groundwater. Earthquake faults can also be entry routes for downward contaminant entry. Water table conditions are of great importance for drinking water supplies, agricultural irrigation, waste disposal (including nuclear waste), wildlife habitat, and other ecological issues.

Interactions with Surface Water

Although interrelated, surface water and groundwater have often been studied and managed as separate resources. Surface water seeps through the soil and becomes groundwater. Conversely, groundwater can also feed surface water sources. Sources of surface water pollution are generally grouped into two categories based on their origin.

Interactions between groundwater and surface water are complex. Consequently, groundwater pollution, sometimes referred to as groundwater contamination, is not as easily classified as surface water pollution. By its very nature, groundwater aquifers are susceptible to contamination from sources that may not directly affect surface water bodies, and the distinction of point vs. non-point source may be irrelevant. A spill or ongoing release of chemical or radionuclide contaminants into soil (located away from a surface water body) may not create point or non-point source pollution but can contaminate the aquifer below, creating a toxic plume.

Prevention

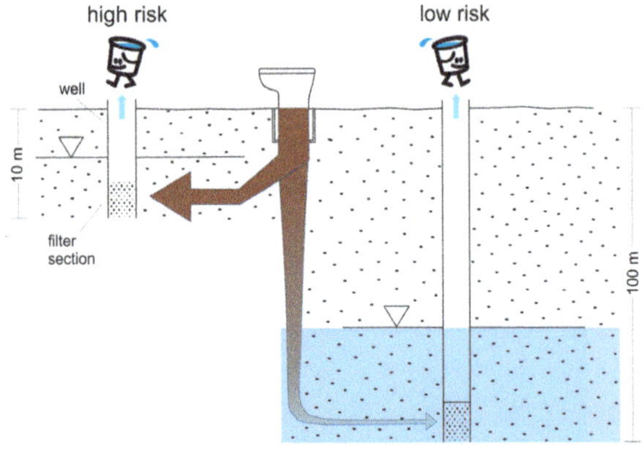

Schematic showing that there is a lower risk of groundwater pollution with greater depth of the water well

Precautionary Principle

The precautionary principle, evolved from Principle 15 of the Rio Declaration on Environment and Development, is important in protecting groundwater resources from pollution. The precautionary principle provides that *"where there are threats of irreversible damage, lack of full scientific certainty shall not be used as reason for postponing cost-effective measures to prevent environmental degradation."*.

One of the six basic principles of the European Union (EU) water policy is the application of the precautionary principle.

Groundwater Quality Monitoring

Groundwater quality monitoring programs have been implemented regularly in many countries around the world. They are important components to understand the hydrogeological system, and for the development of conceptual models and aquifer vulnerability maps.

Groundwater quality must be regularly monitored across the aquifer to determine trends. Effective groundwater monitoring should be driven by a specific objective, for example, a specific contaminant of concern. Contaminant levels can be compared to the World Health Organization (WHO) guidelines for drinking-water quality. It is not rare that limits of contaminants are reduced as more medical experience is gained.

Sufficient investment should be given to continue monitoring over the long term. When a problem is found, action should be taken to correct it. Waterborne outbreaks in the United States decreased with the introduction of more stringent monitoring (and treatment) requirements in the early 90s.

The community can also help monitor the groundwater quality.

Land Zoning for Groundwater Protection

The development of land-use zoning maps has been implemented by several water authorities at different scales around the world. There are two types of zoning maps: aquifer vulnerability maps and source protection maps.

Aquifer Vulnerability Map

It refers to the intrinsic (or natural) vulnerability of a groundwater system to pollution. Intrinsically, some aquifers are more vulnerable to pollution than other aquifers. Shallow unconfined aquifers are more at risk of pollution because there are fewer layers to filter out contaminants.

The unsaturated zone can play an important role in retarding (and in some cases eliminating) pathogens and so must be considered when assessing aquifer vulnerability. The biological activity is greatest in the top soil layers where the attenuation of pathogens is generally most effective.

Preparation of the vulnerability maps typically involves overlaying several thematic maps of physical factors that have been selected to describe the aquifer vulnerability. The index-based parametric mapping method GOD developed by Foster and Hirata (1988) uses three generally available or readily estimated parameters, the degree of Groundwater hydraulic confinement, geological nature of the Overlying strata and Depth to groundwater. A further approach developed by the US EPA named DRASTIC employs seven hydrogeological factors to develop an index of vulnerability: Depth to water table, net Recharge, Aquifer media, Soil media, Topography (slope), Impact on the vadose zone, and hydraulic Conductivity.

There is a particular debate among hydrogeologist whether aquifer vulnerability should be established in a general (intrinsic) way for all contaminants, or specifically for each pollutant.

Source Protection Map

It refers to the capture areas around an individual groundwater source, such as a water well or a spring, to especially protect them from pollution. Thus, potential sources of degradable pollutants, such as pathogens, can be located at distances which travel times along the flowpaths are long enough for the pollutant to be eliminated through filtration or adsorption.

Analytical methods using equations to define groundwater flow and contaminant transport are the most widely used. The WHPA is a semi-analytical groundwater flow simulation program developed by the US EPA for delineating capture zones in a wellhead protection area.

The simplest form of zoning employs fixed-distance methods where activities are excluded within a uniformly applied specified distance around abstraction points.

Locating On-site Sanitation Systems

As the health effects of most toxic chemicals arise after prolonged exposure, risk to health from chemicals is generally lower than that from pathogens. Thus, the quality of the source protection measures is an important component in controlling whether pathogens may be present in the final drinking-water.

On-site sanitation systems can be designed in such a way that groundwater pollution from these sanitation systems is prevented from occurring. Detailed guidelines have been developed to estimate safe distances to protect groundwater sources from pollution from on-site sanitation. The following criteria have been proposed for safe siting (i.e. deciding on the location) of on-site sanitation systems:

- Horizontal distance between the drinking water source and the sanitation system
  - Guideline values for horizontal separation distances between on-site sanitation systems and water sources vary widely (e.g. 15 to 100 m horizontal distance between pit latrine and groundwater wells)
- Vertical distance between drinking water well and sanitation system
- Aquifer type
- Groundwater flow direction
- Impermeable layers
- Slope and surface drainage
- Volume of leaking wastewater
- Superposition, i.e. the need to consider a larger planning area

As a very general guideline it is recommended that the bottom of the pit should be at least 2 m above groundwater level, and a minimum horizontal distance of 30 m between a pit and a water source is normally recommended to limit exposure to microbial contamination. However, no general statement should be made regarding the minimum lateral separation distances required to prevent contamination of a well from a pit latrine. For example, even 50 m lateral separation distance might not be sufficient in a strongly karstified system with a downgradient supply well or spring, while 10 m lateral separation distance is completely sufficient if there is a well developed clay cover layer and the annular space of the groundwater well is well sealed.

Legislation

Sign near Mannheim, Germany indicating a zone as a dedicated "groundwater protection zone"

Institutional and legal issues are critical in determining the success or failure of groundwater protection policies and strategies.

United States

In November 2006, the Environmental Protection Agency published the Ground Water Rule in the United States Federal Register. The EPA was worried that the ground water system would be vulnerable to contamination from fecal matter. The point of the rule was to keep microbial pathogens out of public water sources. The 2006 Ground Water Rule was an amendment of the 1996 Safe Drinking Water Act.

The ways to deal with groundwater pollution that has already occurred can be grouped into the following categories: containing the pollutants to prevent them from migrating further; removing the pollutants from the aquifer; remediating the aquifer by either immobilizing or detoxifying the contaminants while they are still in the aquifer (in-situ); treating the groundwater at its point of use; or abandoning the use of this aquifer's groundwater and finding an alternative source of water.

Management

Point-of-use Treatment

Portable water purification devices or "point-of-use" (POU) water treatment systems and field water disinfection techniques can be used to remove some forms of groundwater pollution prior to drinking, namely any fecal pollution. Many commercial portable water purification systems or chemical additives are available which can remove pathogens, chlorine, bad taste, odors, and heavy metals like lead and mercury.

Techniques include boiling, filtration, activated charcoal absorption, chemical disinfection, ultraviolet purification, ozone water disinfection, solar water disinfection, solar distillation, homemade water filters.

Arsenic removal filters (ARF) are dedicated technologies typically installed to remove arsenic. Many of these technologies require a capital investment and long-term maintenance. Filters in Bangladesh are usually abandoned by the users due to their high cost and complicated maintenance, which is also quite expensive.

Groundwater Remediation

Groundwater pollution is much more difficult to abate than surface pollution because groundwater can move great distances through unseen aquifers. Non-porous aquifers such as clays partially purify water of bacteria by simple filtration (adsorption and absorption), dilution, and, in some cases, chemical reactions and biological activity; however, in some cases, the pollutants merely transform to soil contaminants. Groundwater that moves through open fractures and caverns is not filtered and can be transported as easily as surface water. In fact, this can be aggravated by the human tendency to use natural sinkholes as dumps in areas of karst topography.

Pollutants and contaminants can be removed from ground water by applying various techniques thereby making it safe for use. Ground water treatment (or remediation) techniques span biological, chemical, and physical treatment technologies. Most ground water treatment techniques utilize a combination of technologies. Some of the biological treatment techniques include bioaugmentation, bioventing, biosparging, bioslurping, and phytoremediation. Some chemical treatment techniques include ozone and oxygen gas injection, chemical precipitation, membrane separation, ion exchange, carbon absorption, aqueous chemical oxidation, and surfactant enhanced recovery. Some chemical techniques may be implemented using nanomaterials. Physical treatment techniques include, but are not limited to, pump and treat, air sparging, and dual phase extraction.

Abandonment

If treatment or remediation of the polluted groundwater is deemed to be too difficult or expensive then abandoning the use of this aquifer's groundwater and finding an alternative source of water is the only other option.

Thermal Pollution

The Brayton Point Power Station in Massachusetts discharges heated water to Mount Hope Bay.

Thermal pollution is the degradation of water quality by any process that changes ambient water temperature. A common cause of thermal pollution is the use of water as a coolant by power plants and industrial manufacturers. When water used as a coolant is returned to the natural environment at a higher temperature, the change in temperature decreases oxygen supply and affects ecosystem composition. Fish and other organisms adapted to particular temperature range can be killed by an abrupt change in water temperature (either a rapid increase or decrease) known as "thermal shock."

Urban runoff—stormwater discharged to surface waters from roads and parking lots—can also be a source of elevated water temperatures.

Ecological Effects

Potrero Generating Station discharged heated water into San Francisco Bay. The plant was closed in 2011.

Warm Water

Elevated temperature typically decreases the level of dissolved oxygen of water, as gases are less soluble in hotter liquids. This can harm aquatic animals such as fish, amphibians and other aquatic organisms. Thermal pollution may also increase the metabolic rate of aquatic animals, as enzyme activity, resulting in these organisms consuming more food in a shorter time than if their environment were not changed. An increased metabolic rate may result in fewer resources; the more adapted organisms moving in may have an advantage over organisms that are not used to the warmer temperature. As a result, food chains of the old and new environments may be compromised. Some fish species will avoid stream segments or coastal areas adjacent to a thermal discharge. Biodiversity can be decreased as a result.

High temperature limits oxygen dispersion into deeper waters, contributing to anaerobic conditions. This can lead to increased bacteria levels when there is ample food supply. Many aquatic species will fail to reproduce at elevated temperatures.

Primary producers (e.g. plants, cyanobacteria) are affected by warm water because higher water temperature increases plant growth rates, resulting in a shorter lifespan and species overpopulation. This can cause an algae bloom which reduces oxygen levels.

Temperature changes of even one to two degrees Celsius can cause significant changes in organism metabolism and other adverse cellular biology effects. Principal adverse changes can include rendering cell walls less permeable to necessary osmosis, coagulation of cell proteins, and alteration of enzyme metabolism. These cellular level effects can adversely affect mortality and reproduction.

A large increase in temperature can lead to the denaturing of life-supporting enzymes by breaking down hydrogen- and disulphide bonds within the quaternary structure of the enzymes. Decreased enzyme activity in aquatic organisms can cause problems such as the inability to break down lipids, which leads to malnutrition.

In limited cases, warm water has little deleterious effect and may even lead to improved function of the receiving aquatic ecosystem. This phenomenon is seen especially in seasonal waters and is known as thermal enrichment. An extreme case is derived from the aggregational habits of the manatee, which often uses power plant discharge sites during winter. Projections suggest that manatee populations would decline upon the removal of these discharges.

Cold Water

Releases of unnaturally cold water from reservoirs can dramatically change the fish and macroinvertebrate fauna of rivers, and reduce river productivity. In Australia, where many rivers have warmer temperature regimes, native fish species have been eliminated, and macroinvertebrate fauna have been drastically altered. This may be mitigated by designing the dam to release warmer surface waters instead of the colder water at the bottom of the reservoir.

Thermal Shock

When a power plant first opens or shuts down for repair or other causes, fish and other organisms adapted to particular temperature range can be killed by the abrupt change in water temperature, either an increase or decrease, known as "thermal shock.":

Sources and Control of Thermal Pollution

Cooling tower at Gustav Knepper Power Station, Dortmund, Germany

Industrial Wastewater

In the United States, about 75 to 82 percent of thermal pollution is generated by power plants. The remainder is from industrial sources such as petroleum refineries, pulp and paper mills, chemical plants, steel mills and smelters. Heated water from these sources may be controlled with:

- cooling ponds, man-made bodies of water designed for cooling by evaporation, convection, and radiation
- cooling towers, which transfer waste heat to the atmosphere through evaporation and/or heat transfer
- cogeneration, a process where waste heat is recycled for domestic and/or industrial heating purposes.

Some facilities use once-through cooling (OTC) systems which do not reduce temperature as effectively as the above systems. For example, the Potrero Generating Station in San Francisco (closed in 2011), used OTC and discharged water to San Francisco Bay approximately 10 °C (20 °F) above the ambient bay temperature.

A bioretention cell for treating urban runoff in California

Urban Runoff

During warm weather, urban runoff can have significant thermal impacts on small streams, as storm water passes over hot parking lots, roads and sidewalks. Storm water management facilities that absorb runoff or direct it into groundwater, such as bioretention systems and infiltration basins, can reduce these thermal effects. These and related systems for managing runoff are components of an expanding urban design approach commonly called green infrastructure.

Retention basins (stormwater ponds) tend to be less effective at reducing runoff temperature, as the water may be heated by the sun before being discharged to a receiving stream.

Marine Pollution

While marine pollution can be obvious, as with the marine debris shown above, it is often the pollutants that cannot be seen that cause most harm.

Marine pollution occurs when harmful, or potentially harmful, effects result from the entry into the ocean of chemicals, particles, industrial, agricultural and residential waste, noise, or the spread of invasive organisms. Eighty percent of marine pollution comes from land. Air pollution is also a contributing factor by carrying off pesticides or dirt into the ocean. Land and air pollution have proven to be harmful to marine life and its habitats.

The pollution often comes from non point sources such as agricultural runoff, wind-blown debris and dust. Nutrient pollution, a form of water pollution, refers to contamination by excessive inputs of nutrients. It is a primary cause of eutrophication of surface waters, in which excess nutrients, usually nitrogen or phosphorus, stimulate algae growth.

Many potentially toxic chemicals adhere to tiny particles which are then taken up by plankton and benthos animals, most of which are either deposit or filter feeders. In this way, the toxins are concentrated upward within ocean food chains. Many particles combine chemically in a manner highly depletive of oxygen, causing estuaries to become anoxic.

When pesticides are incorporated into the marine ecosystem, they quickly become absorbed into marine food webs. Once in the food webs, these pesticides can cause mutations, as well as diseases, which can be harmful to humans as well as the entire food web.

Toxic metals can also be introduced into marine food webs. These can cause a change to tissue matter, biochemistry, behaviour, reproduction, and suppress growth in marine life. Also, many animal feeds have a high fish meal or fish hydrolysate content. In this way, marine toxins can be transferred to land animals, and appear later in meat and dairy products.

History

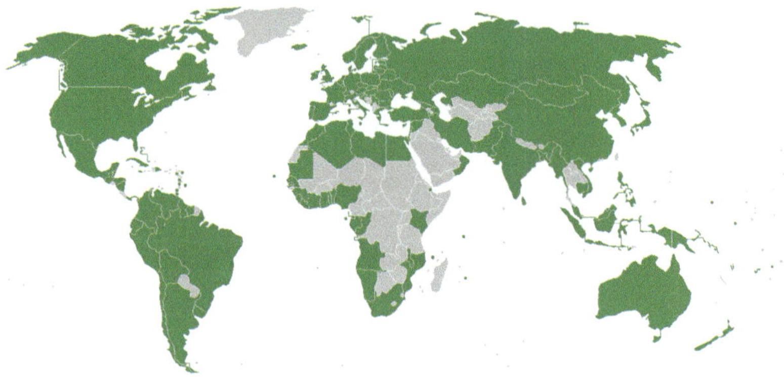

Parties to the MARPOL 73/78 convention on marine pollution

Although marine pollution has a long history, significant international laws to counter it were only enacted in the twentieth century. Marine pollution was a concern during several United Nations Conferences on the Law of the Sea beginning in the 1950s. Most scientists believed that the oceans were so vast that they had unlimited ability to dilute, and thus render pollution harmless.

In the late 1950s and early 1960s, there were several controversies about dumping radioactive waste off the coasts of the United States by companies licensed by the Atomic Energy Commission, into the Irish Sea from the British reprocessing facility at Windscale, and into the Mediterranean Sea by the French Commissariat à l'Energie Atomique. After the Mediterranean Sea controversy, for example, Jacques Cousteau became a worldwide figure in the campaign to stop marine pollution. Marine pollution made further international headlines after the 1967 crash of the oil tanker Torrey Canyon, and after the 1969 Santa Barbara oil spill off the coast of California.

Marine pollution was a major area of discussion during the 1972 United Nations Conference on the Human Environment, held in Stockholm. That year also saw the signing of the Convention on the Prevention of Marine Pollution by Dumping of Wastes and Other Matter, sometimes called the London Convention. The London Convention did not ban marine pollution, but it established black and gray lists for substances to be banned (black) or regulated by national authorities (gray). Cyanide and high-level radioactive waste, for example, were put on the black list. The London Convention applied only to waste dumped from ships, and thus did nothing to regulate waste discharged as liquids from pipelines.

Pathways of Pollution

There are many different ways to categorize, and examine the inputs of pollution into our marine ecosystems. Patin (n.d.) notes that generally there are three main types of inputs of pollution into the ocean: direct discharge of waste into the oceans, runoff into the waters due to rain, and pollutants that are released from the atmosphere.

One common path of entry by contaminants to the sea are rivers. The evaporation of water from oceans exceeds precipitation. The balance is restored by rain over the continents entering rivers and then being returned to the sea. The Hudson in New York State and the Raritan in New Jersey, which empty at the northern and southern ends of Staten Island, are a source of mercury contamination of zooplankton (copepods) in the open ocean. The highest concentration in the filter-feeding copepods is not at the mouths of these rivers but 70 miles south, nearer Atlantic City, because water flows close to the coast. It takes a few days before toxins are taken up by the plankton.

Septic river.

Pollution is often classed as point source or nonpoint source pollution. Point source pollution occurs when there is a single, identifiable, and localized source of the pollution. An example is directly discharging sewage and industrial waste into the ocean. Pollution such as this occurs particularly in developing nations. Nonpoint source pollution occurs when the pollution comes from ill-defined and diffuse sources. These can be difficult to regulate. Agricultural runoff and wind blown debris are prime examples.

Direct Discharge

Acid mine drainage in the Rio Tinto River.

Pollutants enter rivers and the sea directly from urban sewerage and industrial waste discharges, sometimes in the form of hazardous and toxic wastes.

Inland mining for copper, gold. etc., is another source of marine pollution. Most of the pollution is simply soil, which ends up in rivers flowing to the sea. However, some minerals discharged in the course of the mining can cause problems, such as copper, a common industrial pollutant, which can interfere with the life history and development of coral polyps. Mining has a poor environmental track record. For example, according to the United States Environmental Protection Agency, mining has contaminated portions of the headwaters of over 40% of watersheds in the western continental US. Much of this pollution finishes up in the sea.

Land Runoff

Surface runoff from farming, as well as urban runoff and runoff from the construction of roads, buildings, ports, channels, and harbours, can carry soil and particles laden with carbon, nitrogen, phosphorus, and minerals. This nutrient-rich water can cause fleshy algae and phytoplankton to thrive in coastal areas; known as algal blooms, which have the potential to create hypoxic conditions by using all available oxygen.

Polluted runoff from roads and highways can be a significant source of water pollution in coastal areas. About 75% of the toxic chemicals that flow into Puget Sound are carried by stormwater that runs off paved roads and driveways, rooftops, yards and other developed land.

Ship Pollution

A cargo ship pumps ballast water over the side.

Ships can pollute waterways and oceans in many ways. Oil spills can have devastating effects. While being toxic to marine life, polycyclic aromatic hydrocarbons (PAHs), found in crude oil, are very difficult to clean up, and last for years in the sediment and marine environment.

Oil spills are probably the most emotive of marine pollution events. However, while a tanker wreck may result in extensive newspaper headlines, much of the oil in the world's seas comes from other smaller sources, such as tankers discharging ballast water from oil tanks used on return ships, leaking pipelines or engine oil disposed of down sewers.

Discharge of cargo residues from bulk carriers can pollute ports, waterways and oceans. In many instances vessels intentionally discharge illegal wastes despite foreign and domestic regulation prohibiting such actions. It has been estimated that container ships lose over 10,000 containers at sea each year (usually during storms). Ships also create noise pollution that disturbs natural wildlife, and water from ballast tanks can spread harmful algae and other invasive species.

Ballast water taken up at sea and released in port is a major source of unwanted exotic marine life. The invasive freshwater zebra mussels, native to the Black, Caspian and Azov seas, were probably transported to the Great Lakes via ballast water from a transoceanic vessel. Meinesz believes that one of the worst cases of a single invasive species causing harm to an ecosystem can be attributed to a seemingly harmless jellyfish. *Mnemiopsis leidyi*, a species of comb jellyfish that spread so it now inhabits estuaries in many parts of the world. It was first introduced in 1982, and thought to have been transported to the Black Sea in a ship's ballast water. The population of the jellyfish shot up exponentially and, by 1988, it was wreaking havoc upon the local fishing industry. "The anchovy catch fell from 204,000 tons in 1984 to 200 tons in 1993; sprat from 24,600 tons in 1984 to 12,000 tons in 1993; horse mackerel from 4,000 tons in 1984 to zero in 1993." Now that the jellyfish have exhausted the zooplankton, including fish larvae, their numbers have fallen dramatically, yet they continue to maintain a stranglehold on the ecosystem.

Invasive species can take over once occupied areas, facilitate the spread of new diseases, introduce new genetic material, alter underwater seascapes and jeopardize the ability of native species to obtain food. Invasive species are responsible for about $138 billion annually in lost revenue and management costs in the US alone.

Atmospheric Pollution

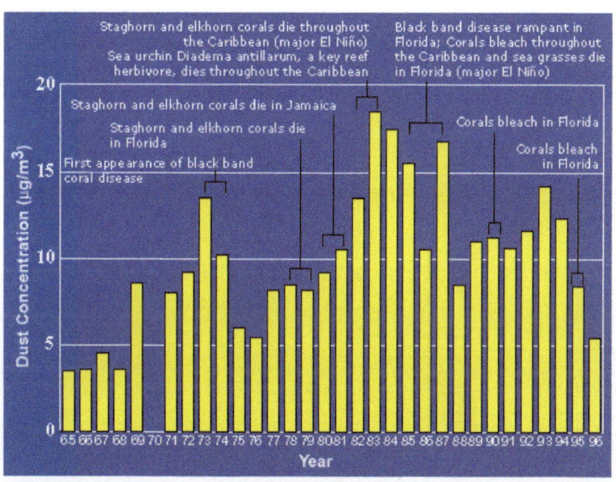

Graph linking atmospheric dust to various coral deaths across the Caribbean Sea and Florida

Another pathway of pollution occurs through the atmosphere. Wind blown dust and debris, including plastic bags, are blown seaward from landfills and other areas. Dust from the Sahara moving around the southern periphery of the subtropical ridge moves into the Caribbean and Florida during the warm season as the ridge builds and moves northward through the subtropical Atlantic. Dust can also be attributed to a global transport from the Gobi and Taklamakan deserts across Korea, Japan, and the Northern Pacific to the Hawaiian Islands. Since 1970, dust outbreaks have worsened due to periods of drought in Africa. There is a large variability in dust transport to the Caribbean and Florida from year to year; however, the flux is greater during positive phases of the North Atlantic Oscillation. The USGS links dust events to a decline in the health of coral reefs across the Caribbean and Florida, primarily since the 1970s.

Climate change is raising ocean temperatures and raising levels of carbon dioxide in the atmosphere. These rising levels of carbon dioxide are acidifying the oceans. This, in turn, is altering aquatic ecosystems and modifying fish distributions, with impacts on the sustainability of fisheries and the livelihoods of the communities that depend on them. Healthy ocean ecosystems are also important for the mitigation of climate change.

Deep Sea Mining

Deep sea mining is a relatively new mineral retrieval process that takes place on the ocean floor. Ocean mining sites are usually around large areas of polymetallic nodules or active and extinct hydrothermal vents at about 1,400 – 3,700 meters below the ocean's surface. The vents create sulfide deposits, which contain precious metals such as silver, gold, copper, manganese, cobalt, and zinc. The deposits are mined using either hydraulic pumps or bucket systems that take ore to the surface to be processed. As with all mining operations, deep sea mining raises questions about environmental damages to the surrounding areas

Because deep sea mining is a relatively new field, the complete consequences of full scale mining operations are unknown. However, experts are certain that removal of parts of the sea floor will result in disturbances to the benthic layer, increased toxicity of the water column and sediment plumes from tailings. Removing parts of the sea floor disturbs the habitat of benthic organisms, possibly, depending on the type of mining and location, causing permanent disturbances. Aside from direct impact of mining the area, leakage, spills and corrosion would alter the mining area's chemical makeup.

Among the impacts of deep sea mining, sediment plumes could have the greatest impact. Plumes are caused when the tailings from mining (usually fine particles) are dumped back into the ocean, creating a cloud of particles floating in the water. Two types of plumes occur: near bottom plumes and surface plumes. Near bottom plumes occur when the tailings are pumped back down to the mining site. The floating particles increase the turbidity, or cloudiness, of the water, clogging filter-feeding apparatuses used by benthic organisms. Surface plumes cause a more serious problem. Depending on the size of the particles and water currents the plumes could spread over vast areas. The plumes could impact zooplankton and light penetration, in turn affecting the food web of the area.

Types of Pollution

Acidification

Island with fringing reef in the Maldives. Coral reefs are dying around the world.

The oceans are normally a natural carbon sink, absorbing carbon dioxide from the atmosphere. Because the levels of atmospheric carbon dioxide are increasing, the oceans are becoming more acidic. The potential consequences of ocean acidification are not fully understood, but there are concerns that structures made of calcium carbonate may become vulnerable to dissolution, affecting corals and the ability of shellfish to form shells.

Oceans and coastal ecosystems play an important role in the global carbon cycle and have removed about 25% of the carbon dioxide emitted by human activities between 2000 and 2007 and about half the anthropogenic CO_2 released since the start of the industrial revolution. Rising ocean temperatures and ocean acidification means that the capacity of the ocean carbon sink will gradually get weaker, giving rise to global concerns expressed in the Monaco and Manado Declarations.

A report from NOAA scientists published in the journal Science in May 2008 found that large amounts of relatively acidified water are upwelling to within four miles of the Pacific continental shelf area of North America. This area is a critical zone where most local marine life lives or is born. While the paper dealt only with areas from Vancouver to northern California, other continental shelf areas may be experiencing similar effects.

A related issue is the methane clathrate reservoirs found under sediments on the ocean floors. These trap large amounts of the greenhouse gas methane, which ocean warming has the potential to release. In 2004 the global inventory of ocean methane clathrates was estimated to occupy between one and five million cubic kilometres. If all these clathrates were to be spread uniformly across the ocean floor, this would translate to a thickness between three and fourteen metres. This estimate corresponds to 500–2500 gigatonnes carbon (Gt C), and can be compared with the 5000 Gt C estimated for all other fossil fuel reserves.

Eutrophication

Polluted lagoon.

Effect of eutrophication on marine benthic life

Eutrophication is an increase in chemical nutrients, typically compounds containing nitrogen or phosphorus, in an ecosystem. It can result in an increase in the ecosystem's primary productivity (excessive plant growth and decay), and further effects including lack of oxygen and severe reductions in water quality, fish, and other animal populations.

The biggest culprit are rivers that empty into the ocean, and with it the many chemicals used as fertilizers in agriculture as well as waste from livestock and humans. An excess of oxygen depleting chemicals in the water can lead to hypoxia and the creation of a dead zone.

Estuaries tend to be naturally eutrophic because land-derived nutrients are concentrated where runoff enters the marine environment in a confined channel. The World Resources Institute has identified 375 hypoxic coastal zones around the world, concentrated in coastal areas in Western Europe, the Eastern and Southern coasts of the US, and East Asia, particularly in Japan. In the ocean, there are frequent red tide algae blooms that kill fish and marine mammals and cause respiratory problems in humans and some domestic animals when the blooms reach close to shore.

In addition to land runoff, atmospheric anthropogenic fixed nitrogen can enter the open ocean. A study in 2008 found that this could account for around one third of the ocean's external (non-recycled) nitrogen supply and up to three per cent of the annual new marine biological production. It has been suggested that accumulating reactive nitrogen in the environment may have consequences as serious as putting carbon dioxide in the atmosphere.

One proposed solution to eutrophication in estuaries is to restore shellfish populations, such as oysters. Oyster reefs remove nitrogen from the water column and filter out suspended solids, subsequently reducing the likelihood or extent of harmful algal blooms or anoxic conditions. Filter feeding activity is considered beneficial to water quality by controlling phytoplankton density and sequestering nutrients, which can be removed from the system through shellfish harvest, buried in the sediments, or lost through denitrification. Foundational work toward the idea of improving marine water quality through shellfish cultivation to was conducted by Odd Lindahl et al., using mussels in Sweden.

Plastic Debris

A mute swan builds a nest using plastic garbage.

Marine debris is mainly discarded human rubbish which floats on, or is suspended in the ocean. Eighty percent of marine debris is plastic – a component that has been rapidly accumulating since the end of World War II. The mass of plastic in the oceans may be as high as 100,000,000 tonnes (98,000,000 long tons; 110,000,000 short tons).

Discarded plastic bags, six pack rings and other forms of plastic waste which finish up in the ocean present dangers to wildlife and fisheries. Aquatic life can be threatened through entanglement, suffocation, and ingestion. Fishing nets, usually made of plastic, can be left or lost in the ocean by fishermen. Known as ghost nets, these entangle fish, dolphins, sea turtles, sharks, dugongs, crocodiles, seabirds, crabs, and other creatures, restricting movement, causing starvation, laceration and infection, and, in those that need to return to the surface to breathe, suffocation.

Remains of an albatross containing ingested flotsam

Many animals that live on or in the sea consume flotsam by mistake, as it often looks similar to their natural prey. Plastic debris, when bulky or tangled, is difficult to pass, and may become permanently lodged in the digestive tracts of these animals. Especially when evolutionary adaptions make it impossible for the likes of turtles to reject plastic bags, which resemble jellyfish when immersed in water, as they have a system in their throat to stop slippery foods from otherwise escaping. Thereby blocking the passage of food and causing death through starvation or infection.

Plastics accumulate because they don't biodegrade in the way many other substances do. They will photodegrade on exposure to the sun, but they do so properly only under dry conditions, and water inhibits this process. In marine environments, photodegraded plastic disintegrates into ever smaller pieces while remaining polymers, even down to the molecular level. When floating plastic particles photodegrade down to zooplankton sizes, jellyfish attempt to consume them, and in this way the plastic enters the ocean food chain. Many of these long-lasting pieces end up in the stomachs of marine birds and animals, including sea turtles, and black-footed albatross.

Marine debris on Kamilo Beach, Hawaii, washed up from the Great Pacific Garbage Patch

Plastic debris tends to accumulate at the centre of ocean gyres. In particular, the Great Pacific Garbage Patch has a very high level of plastic particulate suspended in the upper water column. In samples taken in 1999, the mass of plastic exceeded that of zooplankton (the dominant animal life in the area) by a factor of six. Midway Atoll, in common with all the Hawaiian Islands, receives substantial amounts of debris from the garbage patch. Ninety percent plastic, this debris accumulates on the beaches of Midway where it becomes a hazard to the bird population of the island.

Midway Atoll is home to two-thirds (1.5 million) of the global population of Laysan albatross. Nearly all of these albatross have plastic in their digestive system and one-third of their chicks die.

Toxic additives used in the manufacture of plastic materials can leach out into their surroundings when exposed to water. Waterborne hydrophobic pollutants collect and magnify on the surface of plastic debris, thus making plastic far more deadly in the ocean than it would be on land. Hydrophobic contaminants are also known to bioaccumulate in fatty tissues, biomagnifying up the food chain and putting pressure on apex predators. Some plastic additives are known to disrupt the endocrine system when consumed, others can suppress the immune system or decrease reproductive rates. Floating debris can also absorb persistent organic pollutants from seawater, including PCBs, DDT and PAHs. Aside from toxic effects, when ingested some of these are mistaken by the animal brain for estradiol, causing hormone disruption in the affected wildlife.

Toxins

Apart from plastics, there are particular problems with other toxins that do not disintegrate rapidly in the marine environment. Examples of persistent toxins are PCBs, DDT, TBT, pesticides, furans, dioxins, phenols and radioactive waste. Heavy metals are metallic chemical elements that have a relatively high density and are toxic or poisonous at low concentrations. Examples are mercury, lead, nickel, arsenic and cadmium. Such toxins can accumulate in the tissues of many species of aquatic life in a process called bioaccumulation. They are also known to accumulate in benthic environments, such as estuaries and bay muds: a geological record of human activities of the last century.

Specific examples

- Chinese and Russian industrial pollution such as phenols and heavy metals in the Amur River have devastated fish stocks and damaged its estuary soil.
- Wabamun Lake in Alberta, Canada, once the best whitefish lake in the area, now has unacceptable levels of heavy metals in its sediment and fish.
- Acute and chronic pollution events have been shown to impact southern California kelp forests, though the intensity of the impact seems to depend on both the nature of the contaminants and duration of exposure.
- Due to their high position in the food chain and the subsequent accumulation of heavy metals from their diet, mercury levels can be high in larger species such as bluefin and albacore. As a result, in March 2004 the United States FDA issued guidelines recommending that pregnant women, nursing mothers and children limit their intake of tuna and other types of predatory fish.
- Some shellfish and crabs can survive polluted environments, accumulating heavy metals or toxins in their tissues. For example, mitten crabs have a remarkable ability to survive in highly modified aquatic habitats, including polluted waters. The farming and harvesting of such species needs careful management if they are to be used as a food.

- Surface runoff of pesticides can alter the gender of fish species genetically, transforming male into female fish.
- Heavy metals enter the environment through oil spills – such as the Prestige oil spill on the Galician coast – or from other natural or anthropogenic sources.
- In 2005, the 'Ndrangheta, an Italian mafia syndicate, was accused of sinking at least 30 ships loaded with toxic waste, much of it radioactive. This has led to widespread investigations into radioactive-waste disposal rackets.
- Since the end of World War II, various nations, including the Soviet Union, the United Kingdom, the United States, and Germany, have disposed of chemical weapons in the Baltic Sea, raising concerns of environmental contamination.

Underwater Noise

Marine life can be susceptible to noise or the sound pollution from sources such as passing ships, oil exploration seismic surveys, and naval low-frequency active sonar. Sound travels more rapidly and over larger distances in the sea than in the atmosphere. Marine animals, such as cetaceans, often have weak eyesight, and live in a world largely defined by acoustic information. This applies also to many deeper sea fish, who live in a world of darkness. Between 1950 and 1975, ambient noise at one location in the Pacific Ocean increased by about ten decibels (that is a tenfold increase in intensity).

Noise also makes species communicate louder, which is called the Lombard vocal response. Whale songs are longer when submarine-detectors are on. If creatures don't "speak" loud enough, their voice can be masked by anthropogenic sounds. These unheard voices might be warnings, finding of prey, or preparations of net-bubbling. When one species begins speaking louder, it will mask other species voices, causing the whole ecosystem to eventually speak louder.

According to the oceanographer Sylvia Earle, "Undersea noise pollution is like the death of a thousand cuts. Each sound in itself may not be a matter of critical concern, but taken all together, the noise from shipping, seismic surveys, and military activity is creating a totally different environment than existed even 50 years ago. That high level of noise is bound to have a hard, sweeping impact on life in the sea."

Adaptation and Mitigation

Much anthropogenic pollution ends up in the ocean. The 2011 edition of the United Nations Environment Programme Year Book identifies as the main emerging environmental issues the loss to the oceans of massive amounts of phosphorus, "a valuable fertilizer needed to feed a growing global population", and the impact billions of pieces of plastic waste are having globally on the health of marine environments. Bjorn Jennssen (2003) notes in his article, "Anthropogenic pollution may reduce biodiversity and productivity of marine ecosystems, resulting in reduction and depletion of human marine food resources". There are two ways the overall level of this pollution can be mitigated: either the human population is reduced, or a way is found to reduce the ecological footprint left behind by the average human. If the second way is not adopted, then the first way may be imposed as world ecosystems falter.

The second way is for humans, individually, to pollute less. That requires social and political will, together with a shift in awareness so more people respect the environment and are less disposed to abuse it. At an operational level, regulations, and international government participation is needed. It is often very difficult to regulate marine pollution because pollution spreads over international barriers, thus making regulations hard to create as well as enforce.

Aerosol can polluting a beach

Without appropriate awareness of marine pollution, the necessary global will to effectively address the issues may prove inadequate. Balanced information on the sources and harmful effects of marine pollution need to become part of general public awareness, and ongoing research is required to fully establish, and keep current, the scope of the issues. As expressed in Daoji and Dag's research, one of the reasons why environmental concern is lacking among the Chinese is because the public awareness is low and therefore should be targeted.

Eutrophication

Effects of Eutrophication

Water Pollution

- Pollution of water because of addition of poisonous substances into water
- Types of water pollution

1. Pollution by Point Sources: Contamination of water bodies by substances released from a single, identifiable source such as a pipe- Most common example is the industrial pollution where a lot many harmful substances are released into water bodies.
2. Pollution by nonPoint Sources: Contamination of water bodies by substances that do not come from a discrete, specified source.

- Pollution by water from farmlands that mixes in streams is a common example for this.
- Another kind of pollution is ground water pollution when polluted water at the surface sinks into the earth to form ground water.

Point source pollution-Pollutants mixing in water bodies

A pattern of pollutants getting mixed in a water body

Waste water from farms contaminates water bodies

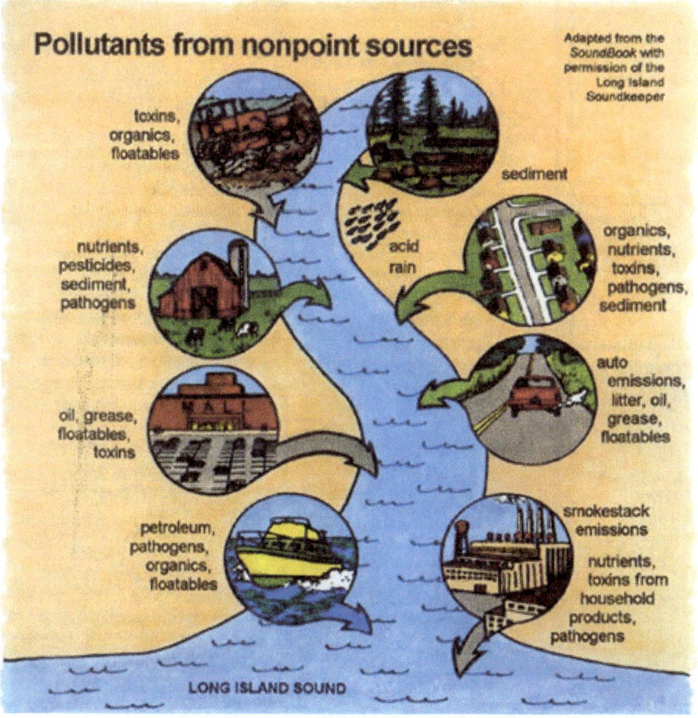

Sources of Nonpoint source pollution

Eutrophication

- Eutrophication is the process by which a body of water acquires a high concentration of nutrients, especially phosphates and nitrates. These typically promote excessive growth of algae which is called "Algal Bloom". As the algae die and decompose, high levels of organic matter and the decomposing organisms deplete the water of available oxygen, causing the death of other organisms, such as fish.
- Eutrophication can occur naturally. It is considered as a natural ageing process of the water body but human interference has paced up the degradation process
- The natural process is called "Natural Eutrophication" and Eutrophication by human in-

terference where large amounts of pollutants are released into water because of societies can be termed as "Cultural Eutrophication"

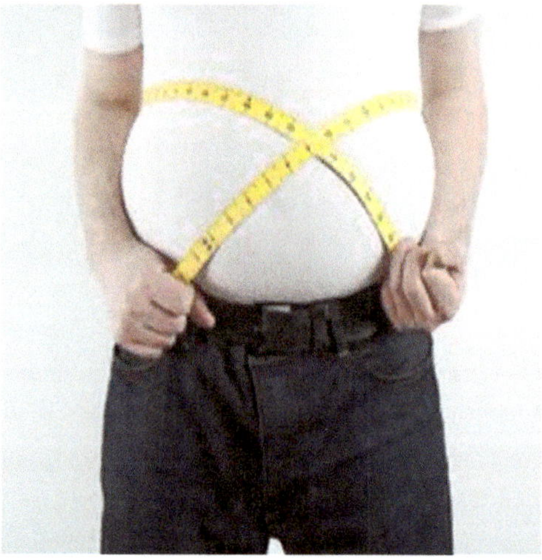

Obesity to humans ~ Eutrophication to water bodies

Picture showing the excess growth of plants in water because of excess release of nutrients into it

Eutrophication, naturally is both prominent in fresh water bodies like lakes & rivers and large salt water bodies like seas & oceans.

1. Lakes and rivers generally experience Eutrophication because of increased phosphorous levels and Seas & oceans because of increased Nitrogen levels.
2. Lakes: Classified based on the nutrient content present in them and production of organic matter:
   - Oligotrophic- Less nutrients- In low nutrient medium, life is also less
   - Mesotrophic - Not many nutrients and not very less nutrients
   - Eutrophic- High nutrients- some lakes naturally have higher contents of nutrients
3. In some water bodies, waste released from the living entities present inside them, rain water, atmospheric nitrogen etc., add to the excess nutrients that constitute Eutrophication.

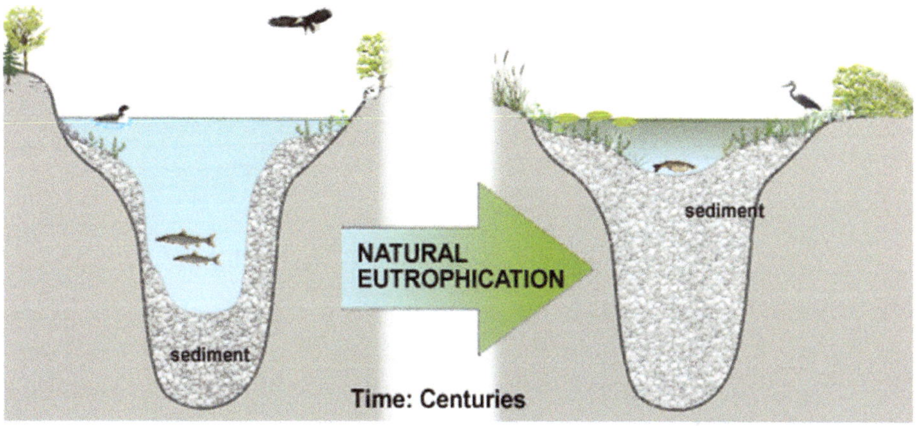

4. Takes a long time for the water body to get polluted to a considerable extent.
5. Factors like availability of light and other resources too influence the growth of unnecessary algae and other organisms that are responsible for Eutrophication-only if it is a long natural process.
6. But because of Human interference and the extent of pollution that is going on now, it would take less than a century for even a large body to get succumbed to Eutrophication.

Cultural Eutrophication

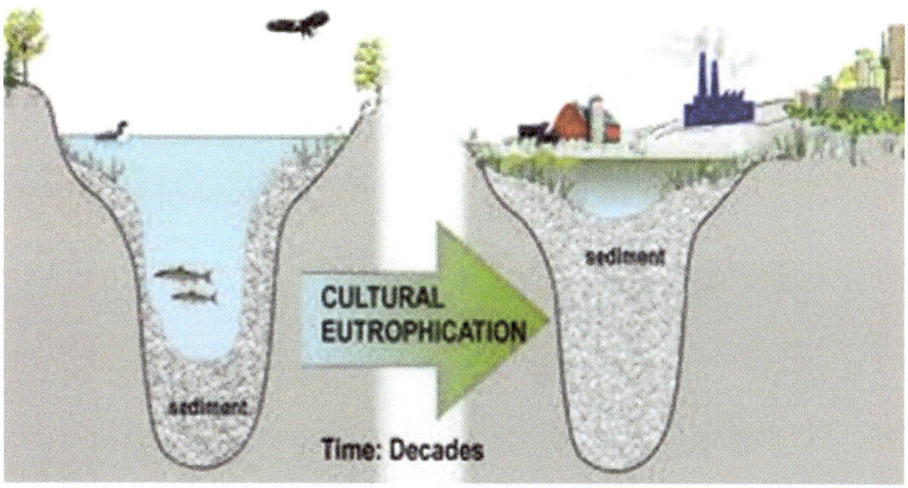

1. Release of pollutants into water which increase the nutrient concentration.
2. Many ways:
   - Sewage disposal in water
   - Industrial waste disposal in water
   - Fertilizers and pesticides dissolved in water being channelized into water bodies
   - Organic fertilizers too can cause Eutrophication
   - Dissolution of pollutant particles on surface in rain water and flowing into water bodies

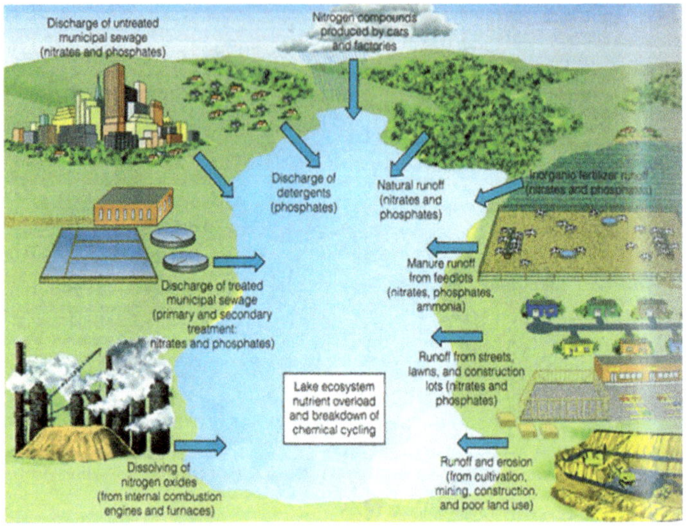

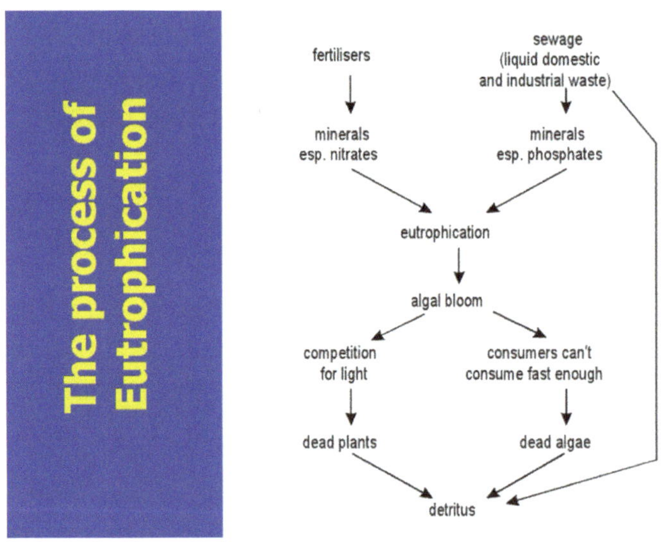

Steps in the Process

1. The process of Eutrophication can be roughly divided into the following steps based on the progress of process:
   - Addition of nutrients
   - Growth of plants/algae
   - Death of plants/algae
   - Growth of bacteria
   - Lack of Oxygen and other unfavourable conditions for existing living entities
   - Suffocation and death of living entities, Sedimentation because of excess waste(Detritus)
   - Death of the water body and eventual drying of land

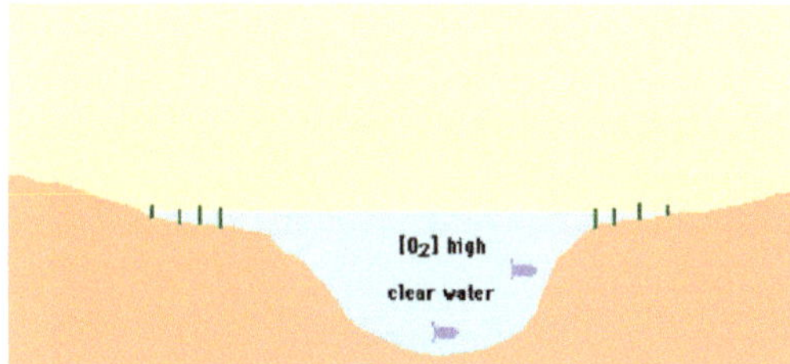

Oligotrophic lake with a low level of nutrients.

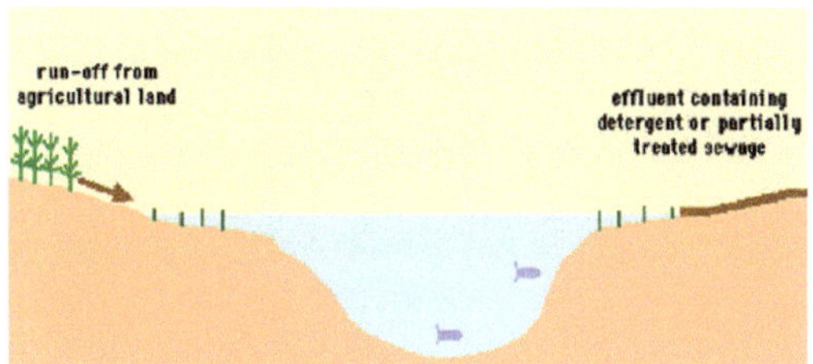

Artificial input of nutrients from run-off and discharge of effluent.

Excess nutrients get added

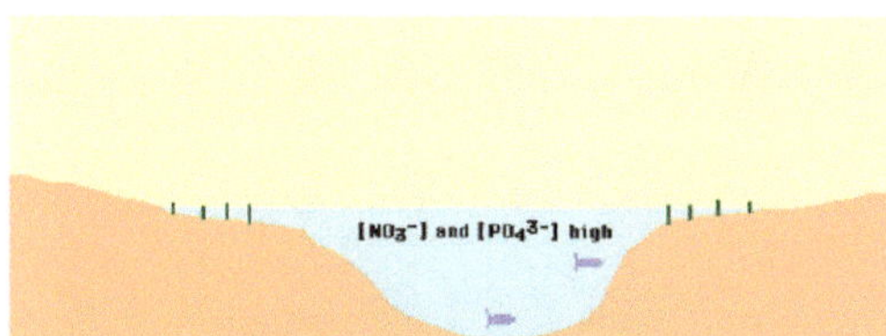

Eutrophic lake with a high level of nutrients. Phosphorus is usually the bio-limiting element in freshwater lakes.

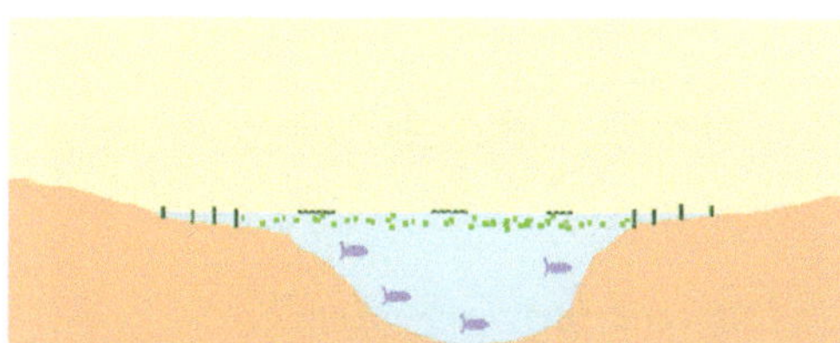

Rapid growth of algae and other biomass resulting in a decrease in the concentration of dissolved oxygen.

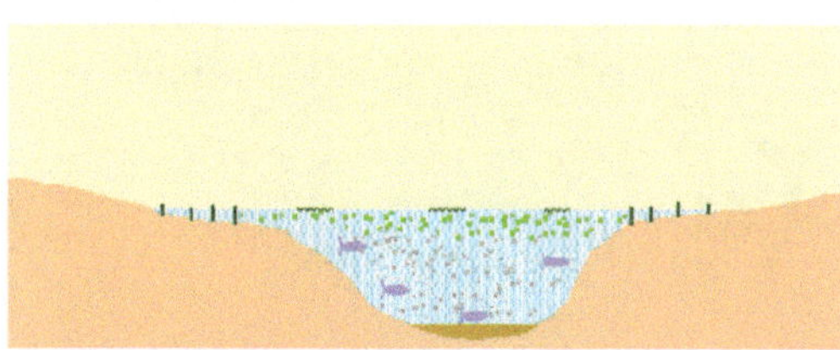

Turbidity (cloudiness) of water increases as does rate of sedimentation.

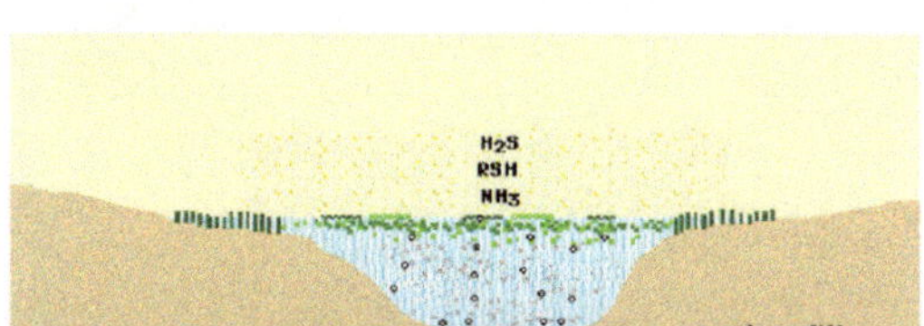

Development of anoxic conditions and release of noxious gases such as hydrogen sulphide, thioalcohols and ammonia.

Effects

- Because of algal bloom, sunlight cannot enter into the depths of the water body which makes plants unable to live ➔ Decrease in O_2 levels occurs resulting in the eventual death of living organisms inside
- During this process of death of living organisms inside, more oxygen is consumed during the decomposition process
- Eutrophication not only effects the living organisms inside the water but also degrades the water quality: Water taste and odor degrade because of the toxins like Geosmin , MIB etc., being released during algal bloom

Before increases in nutrients:

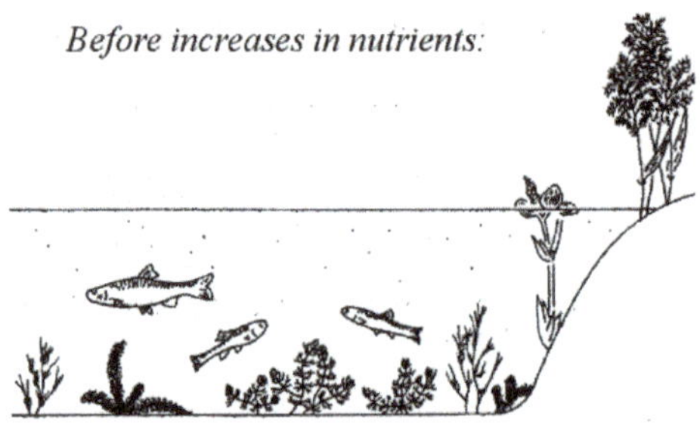

Acceleration of eutrophication :

Results of eutrophication :

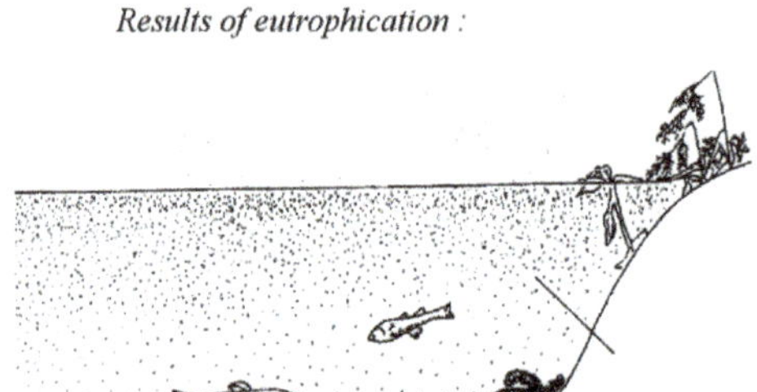

- In lakes and rivers, Eutrophication triggers increase in water level in the long run which creates a problem for shore plant growth and their survival.
- This not only contaminates the water body but eventually ends the plant life on the shore too!

- As water bodies go on dying, the food chain breaks. Only those species which take high nutrients and can grab them fast survive in the habitat
- Tubificid worms, larvae and bacteria replace the vertebrates in water
- This broken food chain can lead to extinction of many species and many adverse conditions

More effects:

- Increased biomass like phytoplankton in water
- Decrease in harvestable fish and shell fish
- Increase in turbidity of water
- Loss of aesthetic value of water body
- Biological Oxygen Demand(BOD) serves as a measure for finding out about Eutrophication in water
- Amount of O_2 consumption by living organisms in water- more consumption ➔ more organic matter ➔ more nutrients ➔ Eutrophication
- A high BOD indicates occurrence of Eutrophication in the water body

Prevention and Reversal of Eutrophication

- If Organic material addition stops, the bacteria use up all the nutrients available and stop die off after sometime → this can let the left off plants and vertebrates grow their numbers again
- Concentration of dissolved O_2 increases through natural means like atmospheric replenishment and photosynthesis
- Cleaning up of water bodies will help in removing sedimentation and allowing plants and fish grow again
- Organic farming reduces the amount of Nitrites that flow into water when compared to its counterpart of farming

  And

- Controlling pollution and flow of polluting substances into water bodies

Acid Mine Drainage

Yellow boy in a stream receiving acid drainage from surface coal mining.

Rocks stained by acid mine drainage on Shamokin Creek

Acid mine drainage, acid and metalliferous drainage (AMD), or acid rock drainage (ARD) refers to the outflow of acidic water from metal mines or coal mines.

Acid rock drainage occurs naturally within some environments as part of the rock weathering process but is exacerbated by large-scale earth disturbances characteristic of mining and other large construction activities, usually within rocks containing an abundance of sulfide minerals. Areas where the earth has been disturbed (e.g. construction sites, subdivisions, and transportation corridors) may create acid rock drainage. In many localities, the liquid that drains from coal stocks, coal handling facilities, coal washeries, and coal waste tips can be highly acidic, and in such cases it is treated as acid rock drainage.

The same type of chemical reactions and processes may occur through the disturbance of acid sulfate soils formed under coastal or estuarine conditions after the last major sea level rise, and constitutes a similar environmental hazard.

Nomenclature

Historically, the acidic discharges from active or abandoned mines were called acid mine drainage, or AMD. The term acid rock drainage, or ARD, was introduced in the 1980s and 1990s to indicate that acidic drainage can originate from sources other than mines. For example, a paper presented in 1991 at a major international conference on this subject was titled: "The Prediction of Acid Rock Drainage - Lessons from the Database" Both AMD and ARD refer to low pH or acidic waters caused by the oxidation of sulfide minerals, though ARD is the more generic name.

In cases where drainage from a mine is not acidic and has dissolved metals or metalloids, or was originally acidic, but has been neutralized along its flow path, then it is described as "Neutral Mine Drainage", "Mining-Influenced Water" or otherwise. None of these other names have gained general acceptance.

Occurrence

In this case, the pyrite has dissolved away yielding a cube shape and residual gold. This break down is the main driver of acid mine drainage.

Sub-surface mining often progresses below the water table, so water must be constantly pumped out of the mine in order to prevent flooding. When a mine is abandoned, the pumping ceases, and water floods the mine. This introduction of water is the initial step in most acid rock drainage situations. Tailings piles or ponds, mine waste rock dumps, and coal spoils are also an important source of acid mine drainage.

After being exposed to air and water, oxidation of metal sulfides (often pyrite, which is iron-sulfide) within the surrounding rock and overburden generates acidity. Colonies of bacteria and archaea greatly accelerate the decomposition of metal ions, although the reactions also occur in an abiotic environment. These microbes, called extremophiles for their ability to survive in harsh conditions, occur naturally in the rock, but limited water and oxygen supplies usually keep their numbers low. Special extremophiles known as Acidophiles especially favor the low pH levels of abandoned mines. In particular, *Acidithiobacillus ferrooxidans* is a key contributor to pyrite oxidation.

Metal mines may generate highly acidic discharges where the ore is a sulfide mineral or is associated with pyrite. In these cases the predominant metal ion may not be iron but rather zinc, copper, or nickel. The most commonly mined ore of copper, chalcopyrite, is itself a copper-iron-sulfide and occurs with a range of other sulfides. Thus, copper mines are often major culprits of acid mine drainage.

At some mines, acidic drainage is detected within 2–5 years after mining begins, whereas at other mines, it is not detected for several decades. In addition, acidic drainage may be generated for decades or centuries after it is first detected. For this reason, acid mine drainage is considered a serious long-term environmental problem associated with mining.

Chemistry

The chemistry of oxidation of pyrites, the production of ferrous ions and subsequently ferric ions, is very complex, and this complexity has considerably inhibited the design of effective treatment options.

Although a host of chemical processes contribute to acid mine drainage, pyrite oxidation is by far the greatest contributor. A general equation for this process is:

$$2FeS_2(s) + 7O_2(g) + 2H_2O(l) = 2Fe^{2+}(aq) + 4SO_4^{2-}(aq) + 4H^+(aq)$$

The oxidation of the sulfide to sulfate solubilizes the ferrous iron (iron(II)), which is subsequently oxidized to ferric iron (iron(III)):

$$4Fe^{2+}(aq) + O_2(g) + 4H^+(aq) = 4Fe^{3+}(aq) + 2H_2O(l)$$

Either of these reactions can occur spontaneously or can be catalyzed by microorganisms that derive energy from the oxidation reaction. The ferric cations produced can also oxidize additional pyrite and reduce into ferrous ions:

$$FeS_2(s) + 14Fe^{3+}(aq) + 8H_2O(l) = 15Fe^{2+}(aq) + 2SO_4^{2-}(aq) + 16H^+(aq)$$

The net effect of these reactions is to release H^+, which lowers the pH and maintains the solubility of the ferric ion.

Effects

Effects on pH

Water temperatures as high as 47 °C have been measured underground at the Iron Mountain Mine, and the pH can be as low as -3.6.

Organisms which cause acid mine drainage can thrive in waters with pH very close to zero. Negative pH occurs when water evaporates from already acidic pools thereby increasing the concentration of hydrogen ions.

About half of the coal mine discharges in Pennsylvania have pH under 5. However, a significant portion of mine drainage in both the bituminous and anthracite regions of Pennsylvania is alkaline, because limestone in the overburden neutralizes acid before the drainage emanates.

Rio Tinto in Spain.

Acid rock drainage has recently been a hindrance to the completion of the construction of Interstate 99 near State College, Pennsylvania. However, this acid rock drainage didn't come from a mine; rather, it was produced by oxidation of pyrite-rich rock which was unearthed during a road cut and then used as filler material in the I-99 construction. A similar situation developed at the Halifax airport in Canada. It is from these and similar experiences that the term acid *rock* drainage has emerged as being preferable to acid *mine* drainage, thereby emphasizing the general nature of the problem.

Yellow Boy

When the pH of acid mine drainage is raised past 3, either through contact with fresh water or neutralizing minerals, previously soluble iron(III) ions precipitate as iron(III) hydroxide, a yellow-orange solid colloquially known as *yellow boy*. Other types of iron precipitates are possible, including iron oxides and oxyhydroxides. All these precipitates can discolor water and smother plant and animal life on the streambed, disrupting stream ecosystems (a specific offense under the Fisheries Act in Canada). The process also produces additional hydrogen ions, which can further decrease pH. In some cases, the concentrations of iron hydroxides in yellow boy are so high, the precipitate can be recovered for commercial use in pigments.

Trace Metal and Semi-metal Contamination

Many acid rock discharges also contain elevated levels of potentially toxic metals, especially nickel and copper with lower levels of a range of trace and semi-metal ions such as lead, arsenic, aluminium, and manganese. The elevated levels of heavy metals can only be dissolved in waters that have a low pH, as is found in the acidic waters produced by pyrite oxidation. In the coal belt around the south Wales valleys in the UK highly acidic nickel-rich discharges from coal stocking sites have proved to be particularly troublesome.

Identification and Prediction

In a mining setting it is leading practice to carry out a geochemical assessment of mine materials during the early stages of a project to determine the potential for AMD. The geochemical assessment aims to map the distribution and variability of key geochemical parameters, acid generating and element leaching characteristics.

The assessment may include:

1. Sampling;
2. Static geochemical testwork (e.g. acid-base accounting, sulfur speciation);
3. Kinetic geochemical testwork - Conducting oxygen consumption tests, such as the OxCon, to quantify acidity generation rates
4. Modelling of oxidation, pollutant generation and release; and
5. Modelling of material composition.

Treatment

Oversight

In the United Kingdom, many discharges from abandoned mines are exempt from regulatory control. In such cases the Environment Agency working with partners such as the Coal Authority have provided some innovative solutions, including constructed wetland solutions such as on the River Pelenna in the valley of the River Afan near Port Talbot and the constructed wetland next to the River Neath at Ynysarwed.

Although abandoned underground mines produce most of the acid mine drainage, some recently mined and reclaimed surface mines have produced ARD and have degraded local ground-water and surface-water resources. Acidic water produced at active mines must be neutralized to achieve pH 6-9 before discharge from a mine site to a stream is permitted.

In Canada, work to reduce the effects of acid mine drainage is concentrated under the Mine Environment Neutral Drainage (MEND) program. Total liability from acid rock drainage is estimated to be between $2 billion and $5 billion CAD. Over a period of eight years, MEND claims to have reduced ARD liability by up to $400 million CAD, from an investment of $17.5 million CAD.

Methods

Lime Neutralization

By far, the most commonly used commercial process for treating acid mine drainage is lime precipitation in a high-density sludge (HDS) process. In this application, a slurry of lime is dispersed into a tank containing acid mine drainage and recycled sludge to increase water pH to about 9. At this pH, most toxic metals become insoluble and precipitate, aided by the presence of recycled sludge. Optionally, air may be introduced in this tank to oxidize iron and manganese and assist in their precipitation. The resulting slurry is directed to a sludge-settling vessel, such as a clarifier. In that vessel, clean water will overflow for release, whereas settled metal precipitates (sludge) will be recycled to the acid mine drainage treatment tank, with a sludge-wasting side stream. A number of variations of this process exist, as dictated by the chemistry of ARD, its volume, and other factors. Generally, the products of the HDS process also contain gypsum and unreacted lime, which enhance both its settleability and resistance to re-acidification and metal mobilization.

Less complex variants of this process, such as simple lime neutralization, may involve no more than a lime silo, mixing tank and settling pond. These systems are far less costly to build, but are also less efficient (i.e., longer reaction times are required, and they produce a discharge with higher trace metal concentrations, if present). They would be suitable for relatively small flows or less complex acid mine drainage.

Calcium silicate Neutralization

A calcium silicate feedstock, made from processed steel slag, can also be used to neutralize active acidity in AMD systems by removing free hydrogen ions from the bulk solution, thereby increasing pH. As the silicate anion captures H^+ ions (raising the pH), it forms monosilicic acid (H_4SiO_4), a neutral solute. Monosilicic acid remains in the bulk solution to play many roles in correcting the adverse effects of acidic conditions. In the bulk solution, the silicate anion is very active in neutralizing H^+ cations in the soil solution. While its mode-of-action is quite different from limestone, the ability of calcium silicate to neutralize acid solutions is equivalent to limestone as evidenced by its CCE value of 90-100% and its relative neutralizing value of 98%.

In the presence of heavy metals, calcium silicate reacts in a different manner than limestone. As limestone raises the pH of the bulk solution, and if heavy metals are present, precipitation of the metal hydroxides (with extremely low solubilities) is normally accelerated and the potential of armoring of limestone particles increases significantly. In the calcium silicate aggregate, as silicic acid species are absorbed onto the metal surface, the development of silica layers (mono- and bi-layers) lead to the formation of colloidal complexes with neutral or negative surface charges. These negatively charged colloids create an electrostatic repulsion with each other (as well as with the negatively charged calcium silicate granules) and the sequestered metal colloids are stabilized and remain in a dispersed state - effectively interrupting metal precipitation and reducing vulnerability of the material to armoring.

Carbonate Neutralization

Generally, limestone or other calcareous strata that could neutralize acid are lacking or deficient at sites that produce acidic rock drainage. Limestone chips may be introduced into sites to create a neutralizing effect. Where limestone has been used, such as at Cwm Rheidol in mid Wales, the positive impact has been much less than anticipated because of the creation of an insoluble calcium sulfate layer on the limestone chips, binding the material and preventing further neutralization.

Ion Exchange

Cation exchange processes have previously been investigated as a potential treatment for acid mine drainage. The principle is that an ion exchange resin can remove potentially toxic metals (cationic resins), or chlorides, sulfates and uranyl sulfate complexes (anionic resins) from mine water. Once the contaminants are adsorbed, the exchange sites on resins must be regenerated, which typically requires acidic and basic reagents and generates a brine that contains the pollutants in a concentrated form. A South African company that won the 2013 IChemE (ww.icheme.org) award for water management and supply (treating AMD) have developed a patented ion-exchange process that treats mine effluents (and AMD) economically.

Constructed Wetlands

Constructed wetlands systems have been proposed during the 1980s to treat acid mine drainage generated by the abandoned coal mines in Eastern Appalachia. Generally, the wetlands receive near-neutral water, after it has been neutralized by (typically) a limestone-based treatment process. Metal precipitation occurs from their oxidation at near-neutral pH, complexation with organic matter, precipitation as carbonates or sulfides. The latter results from sediment-borne anaerobic bacteria capable of reverting sulfate ions into sulfide ions. These sulfide ions can then bind with heavy metal ions, precipitating heavy metals out of solution and effectively reversing the entire process.

The attractiveness of a constructed wetlands solution lies in its relative low cost. They are limited by the metal loads they can deal with (either from high flows or metal concentrations), though current practitioners have succeeded in developing constructed wetlands that treat high volumes and/or highly acidic water (with adequate pre-treatment). Typically, the effluent from constructed wetland receiving near-neutral water will be well-buffered at between 6.5-7.0 and can readily be discharged. Some of metal precipitates retained in sediments are unstable when exposed to oxygen (e.g., copper sulfide or elemental selenium), and it is very important that the wetland sediments remain largely or permanently submerged.

An example of an effective constructed wetland is on the Afon Pelena in the River Afan valley above Port Talbot where highly ferruginous discharges from the Whitworth mine have been successfully treated.

Precipitation of Metal Sulfides

Most base metals in acidic solution precipitate in contact with free sulfide, e.g. from H_2S or NaHS. Solid-liquid separation after reaction would produce a base metal-free effluent that can be discharged or further treated to reduce sulfate, and a metal sulfide concentrate with possible economic value.

As an alternative, several researchers have investigated the precipitation of metals using biogenic sulfide. In this process, Sulfate-reducing bacteria oxidize organic matter using sulfate, instead of oxygen. Their metabolic products include bicarbonate, which can neutralize water acidity, and hydrogen sulfide, which forms highly insoluble precipitates with many toxic metals. Although promising, this process has been slow in being adopted for a variety of technical reasons.

Technologies

Many technologies exist for the treatment of AMD from traditional high cost water treatment plants to simple in situ water treatment reagent dosing methods.

Metagenomic Study of Acid Mine Drainage

With the advance of Large-scale sequencing strategies, genomes of microorganisms in the acid mine drainage community are directly sequenced from the environment. The nearly full genomic constructs allows new understanding of the community and able to reconstruct their metabolic pathways. Our knowledge of Acidophiles in acid mine drainage remains rudimentary: we know of many more species associated with ARD than we can establish roles and functions.

Microbes and Drug Discovery

Scientists have recently begun to explore acid mine drainage and mine reclamation sites for unique soil bacteria capable of producing new pharmaceutical leads. Soil microbes have long been a source for effective drugs and new research, such as that conducted at the Center for Pharmaceutical Research and Innovation, suggests these extreme environments to be an untapped source for new discovery.

Acid Mine Drainage

The main cause of acid mine drainage is oxidation of iron pyrite. Iron pyrite, FeS2, is the most widespread of all sulfide minerals and is found in many ore bodies. During mining operations, particularly coal mining, iron pyrite in the ore is exposed to air and water, causing it to be oxidized to sulfuric acid and ferrous ion:

$$FeS_2 + {}^7\!/_2\, O_2 + H_2O \leftrightarrow Fe^{2+} + 2\,SO_4^{2-} + 2\,H^+. \quad (1)$$

There is almost always enough moisture in mine wastes and mine workings to allow above reaction (1) to occur releasing acidity and dissolved ferrous ion into the water. Next, dissolved Fe^{2+} is oxidized slowly by dissolved oxygen to Fe3+, consuming some acidity:

$$Fe^{2+} + {}^1\!/_4\, O_2 + H^+ \leftrightarrow Fe^{3+} + {}^1\!/_2\, H_2O. \quad (2)$$

Above pH 4 and in the absence of iron-oxidizing bacteria, equn (2) is the rate-limiting step in the reaction sequence.

However, below pH 4 and in the presence of iron-oxidizing bacteria, the rate of equn (2) is greatly accelerated by a million-fold or more. Ferric ion formed in equn (2) can further oxidize pyrite, where ferric ion is reduced back to Fe^{2+}, releasing much more acidity.

$$FeS_2(s) + 14\,Fe^{3+} + 18\,H_2O \leftrightarrow 15\,Fe^{2+} + 2\,SO_4^{2-} + 16\,H^+. \quad (3)$$

By equn (3), eight times more acidity is generated when ferric ion oxidizes pyrite than when dissolved oxygen serves as the oxidant (16 equivalents of H^+ compared to 2 equivalents, per mole of FeS_2).

In the pH range from 2 to 7, pyrite oxidation by Fe^{3+} {equn (3)} is kinetically favored over abiotic oxidation by oxygen {equn(2)}.

In addition, equn (3) returns soluble Fe^{2+} to the reaction cycle.

Overall, 4 equivalents of acid are formed for each mole of FeS_2 oxidized in the cyclic reaction sequence of equn (2) and (3).

If bacterially mediated oxidation is occurring in equn (2), the reaction cycle can be accelerated by over a millionfold.

Ferric ion also hydrolyzes (reacts with water), releasing more acid to the water and forming insoluble ferric hydroxide, which can coat streambeds with the yellow-orange deposits known as yellow boy.

$$Fe^{3+} + 3H_2O \leftrightarrow Fe(OH)_3(s) + 3H^+.$$

$Fe(OH)_3$ precipitates serve as a reservoir for dissolved Fe^{3+}.

If the generation of Fe^{3+} by equn (3) is stopped because of lack of oxygen, then Fe^{3+} is supplied by dissolution of solid $Fe(OH)_3$ and is available to react via equn (3).

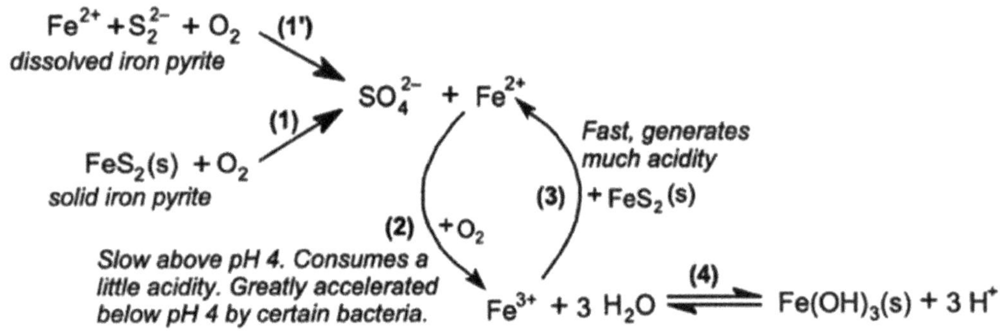

Step 1: Iron pyrite, dissolved or solid, is oxidized by dissolved oxygen equn (1), producing Fe^{3+}, SO_4^{2-}, and lowering the pH.

Step 2: Fe^{2+} formed in Step 1 is oxidized slowly by dissolved oxygen to Fe^{3+}, equn (2). This is the rate-limiting step in the reaction sequence in the absence of iron-oxidizing bacteria. The abiotic rate decreases with lower pH. However, iron-oxidizing bacteria can greatly accelerate this step when the pH falls below 4.

Step 3: Fe^{3+} from Step 2 is reduced rapidly back to Fe^{2+} by pyrite equn (3), generating much acidity. Ferrous ion, Fe^{2+} , generated in Step 3 re-enters the reactioncycle via step 2.

Step 4: A portion of ferric ion, Fe^{3+}, reacts with water to form $Fe(OH)_3$ precipitate, releasing more acidity equn (4).

When theFe^{3+} concentration diminishes, Fe(OH)3 precipitate can dissolve, acting as a reservoir for replenishing Fe^{3+} and maintaining the acid producing cycle.

As pH is lowered, Step 1 becomes less important and the abiotic rate of Step 2 decreases.

However, Step2 can be greatly accelerated by certain bacteria such as Metallogenium, Ferrobacillus,Thiobacillus, and Leptospirillum, which derive energy from the oxidation of Fe^{2+} to Fe^{3+}.

Below pH 4, these bacteria catalyze Step 2, speeding up the overall reaction rate by a factor as large as 106, and can lower the pH to 2 or less.

Furthermore, these bacteria can tolerate high concentrations of dissolved metals (e.g., 40,000 mg/L Zn and Fe; 15,000 mg/L Cu) before experiencing toxic effects.

They thrive in mine drainage waters as long as a minimal amount of oxygen is present. Once bacterial acceleration occurs, it is hard to reverse.

Non Metal Sulphides donot Generate Acidity

The oxidation by dissolved O_2 of noniron metal sulfides does not generate significant amounts of acidity. The metals are released as dissolved cations but acidity is not produced.

$$CuS(s) + 2O_2 \leftrightarrow Cu^{2+}(aq) + SO_4{}^{2-}(aq).$$
$$ZnS(s) + 2O_2 \leftrightarrow Zn^{2+}(aq) + SO_4{}^{2-}(aq).$$
$$PbS(s) + 2O_2 \leftrightarrow Pb^{2+}(aq) + SO_4{}^{2-}(aq).$$

Two possible reasons for the lack of acid formation when noniron sulfides are oxidized are:

$$(iron\ pyrite)\quad S_2^{2-} + {}^{7}\!/_{2}O_2 + H_2O \leftrightarrow 2\,SO_4^{2-} + 2\,H^+$$
$$(other\ sulfides)\quad S^{2-} + 2\,O_2 \leftrightarrow 2\,SO_4^{2-}$$

The oxidation state of sulfur is different in iron pyrite than in other sulfides, occurring as $S_2{}^{2-}$ in iron pyrite and as S^{2-} in other sulfides. The respective oxidation reactions of these two sulfur forms indicate that acid is produced only with $S_2{}^{2-}$.

Cu^{2+}, Zn^{2+}, Pb^{2+}, etc., do not hydrolyze as extensively as does Fe^{3+}, so noniron sulfides do not react significantly by reactions.

$$Fe^{3+} + 3\,H_2O \leftrightarrow Fe(OH)_3(s) + 3\,H^+.$$

Ferric ion, Fe^{3+}, can oxidize other metal sulfides, such as ZnS (sphalerite), CuS (covellite), PbS (galena), and $CuFeS_2$ (chalcopyrite), in a similar fashion to its oxidation of FeS_2, releasing metal cations into the water without generating acidity.

Acid Base Potential of Soil

The acid-base potential (ABP) is a measure of how effectively the alkalinity (neutralization potential) in a solid sample can neutralize the acid-producing potential resulting from the presence of pyrite of the sample. The acid-base potential is equal to the equivalents of calcium carbonate ($CaCO_3$) in excess of the amount needed to neutralize the acid that could potentially be produced from oxidation of pyritic sulfur.

Example: Determining the Acid-Base Potential (ABP)

The ABP is calculated by

ABP = (alkalinity) – (31.25)(wt % pyritic sulfur)

Where, ABP is given in tons of $CaCO_3$-equivalents per 1000 tons of solid material. Any rock or earth material with an ABP of –5.0 represents a soil with a net potential deficiency of 5.0 tons $CaCO_3$/1000 tons material, and is defined as a potentially toxic material.

1. If the ABP is positive, leachate from the sample is likely to be basic.
2. If the ABP is negative, leachate is likely to be acidic.
3. If the ABP is –5, or more negative, the earth material may be defined as a potentially toxic material.

Pesticides and Fertilizers

Pesticides

- Pesticides are substances or mixture of substances intended for preventing, destroying, repelling or mitigating any pest.
- Green Revolution leads to the usage of pesticides, fertilizers herbicides not only on farms but in backyard gardens and on front lawns as well.

- Many of these chemicals were developed in the lab and are petroleum-based.
- These chemicals allowed farmers and gardeners of every stripe to exercise greater control over the plants they want to grow by enriching the immediate environment and warding off pests.
- At the same time as these synthetic chemicals are abs- orbed by soil and run-off into near by water ways causing land and water pollution.

Pesticides and Pollution of the Soil

- pesticides affect soil quality
- Pesticides decrease biodiversity in the soil because they do not just kill the intended pest; they often kill many of the other small organisms present.
- When life in the soil is killed off, the soil quality deteriorates and this has a knock-on effect upon the retention of water.

- When pesticides kill off most of the active soil organisms, the complex interactions which result in good fertility break down.
- Plants depend on millions of bacteria and fungi to bring nutrients to their rootlets. When these cycles are disrupted plants become more dependent upon exact doses of chemical fertilizers at regular intervals.
- We get large but watery vegetables and fruits, which often lack taste and nutrients and may even contain pesticide residues.
- The misuse of pesticides can cause valuable pollinators such as bees and hover-flies to be killed and this in turn can badly affect food crops.
- There are fears that through natural selection, some pests may eventually become quite resistant to pesticides. Farmers may then try increasing amounts used of pesticides, making the problem worse.
- Pesticides are also polluting water and air.

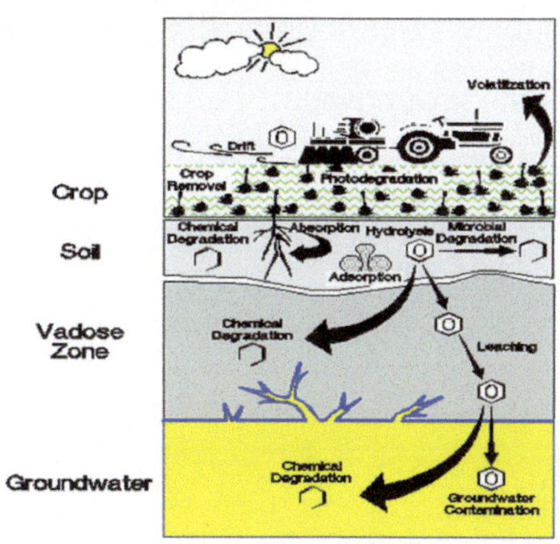

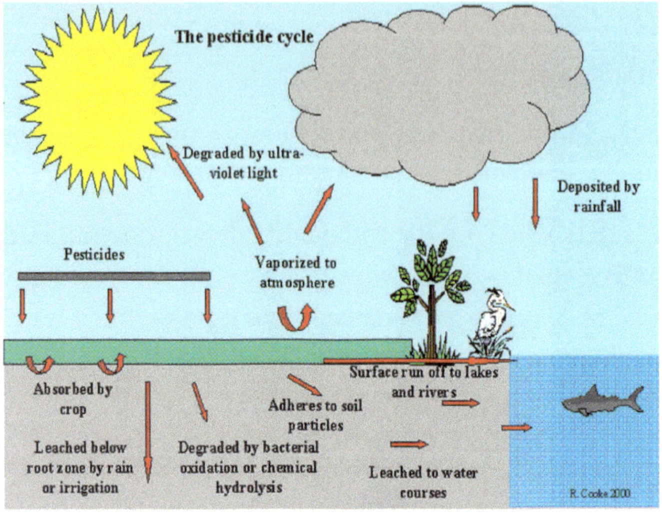

Pesticides and Pollution of the Water

- When the excess nutrients from all the fertilizer and pesticides we use runs off into our waterways, they cause algae blooms sometimes big enough to make waterways impassable.

- When the algae die, they sink to the bottom and decompose in a process that removes oxygen from the water. Fish and other aquatic species can't survive in these so-called "dead zones" and so they die or move on to greener underwater pastures.
- When those chemicals get into waterways, fish ingest them and become diseased. Humans who eat diseased fish can themselves become ill, completing the circle wrought by pollution.
- Pesticides also polluting the groundwater. As 90% of the population use ground water as drinking water it results in great health problems.
- Individual pesticides have unique properties, and many variable factors determine the specific risk in terms of water pollution.

Pesticide half-life

Mobility in soil

Solubility in water

Rainfall

Microbial activity

Soil temperature

Pesticide Handling Practices to Protect Ground Water

- Cleaning a contaminated aquifer is almost impossible. Treatment is complicated, time consuming, and expensive.
- The following pesticide management and handling practices can help you reduce the potential for groundwater contamination.

Practice Integrated Pest Management

Accurately identify the pest and apply pesticides only when necessary, using the lowest label rate that gives adequate pest control. Minimizing pesticide use makes good economic sense and reduces the potential for pest resistance and environmental problems.

Select Pesticides Carefully

Pesticides that are not adsorbed to soil particles can easily move in water, do not vaporize easily, are relatively stable, and thus have the greatest potential to leach into groundwater.

Consider the Vulnerability of the Area

Determine the relative susceptibility of the soil to leaching. Soil texture, organic matter content, and permeability affect pesticide movement. To the extent possible, determine the depth of the water table and the relative permeability of the geologic layers between the soil surface and the groundwater. Avoid spraying ditches and waterways along the field.

Consider the Location and Condition of Wells

Cap and seal wells properly and Slope the surrounding ground away from the wellhead to keep runoff away from the well.

Measure Accurately

Carefully calculate how much pesticide concentrate you need to treat the specific site with the equipment you plan to use. Handle pesticides carefully to avoid spills.

Prevent Back Siphoning

To prevent pesticides from siphoning back into the water supply, keep the end of the fill hose above the water level in the spray tank. Use an anti-backflow device (check valve) on the fill hose, especially when siphoning water directly from a pond or stream.

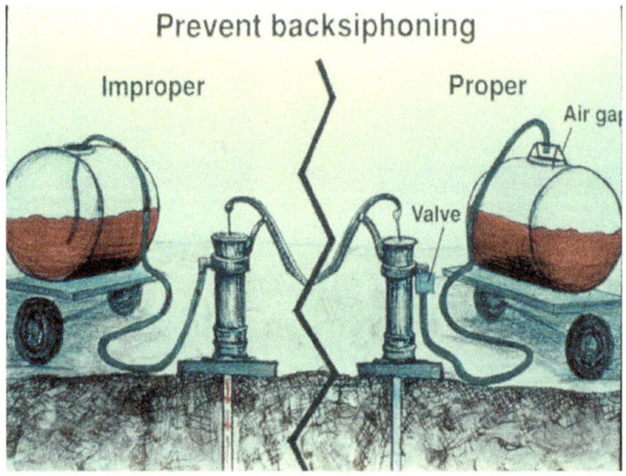

- Delay pesticide application if you expect rain because rainfall will increases pesticide run-off and leaching.
- Construct the pesticide storage area on a concrete floor sealed with an impervious material, located away from all water sources.
- The solution is to use organic materials like composted manure and other natural materials, as well as crop rotation, to help improve soil fertility, rather than synthetic fertilizers and pesticides that can result in soil and water pollution.

Soil & Water Pollution from Fertilizers

To Choose or Not To Choose

What are fertilizers?

- Fertilizer is any organic or inorganic material of natural or synthetic origin added to a soil to supply one or more plant nutrients essential to the growth of plants.
- Fertilizers come in various forms. The most typical form is solid fertilizer in granulated or powdered forms. The next most common form is liquid fertilizer; some advantages of liquid fertilizer are its immediate effect and wide coverage
- Fertilizers are broadly divided into organic fertilizers (composed of organic plant or animal matter), inorganic or commercial fertilizers

- Organic fertilizers include naturally occurring organic materials, (e.g. chicken litter, manure, worm castings, compost, seaweed, guano, bone meal) or naturally occurring mineral deposits (e.g. saltpeter)
- Fertilizers typically provide, in varying proportions:

  six macronutrients: nitrogen (N), phosphorus (P), potassium (K), calcium (Ca), magnesium (Mg), and sulfur (S); eight micronutrients: boron (B), chlorine (Cl), copper (Cu), iron (Fe), manganese (Mn), molybdenum (Mo), zinc (Zn) and nickel (Ni) (1987).
- Inorganic Fertilizers come in various form. Nitrogen fertilizers are often made using the Haber-Bosch process. Phosphate rock can also be processed into water-soluble phosphate (P_2O_5) with the addition of sulphuric acid (H_2SO_4) to make the phosphoric acid in phosphate fertilizers.

Pollution: Importance or Ignorance?

Is it the importance or our ignorance that we have overlooked such an important matter of fertilizers being a major cause of water and soil pollution

Fertilizers have helped to increase crop production and keep pace with the increase in population but their growing use threatens our natural resources like water and soil.

BT Brinjal

Water Pollution

The nutrients, especially nitrates, in fertilizers can cause problems for natural habitats and for human health if they are washed off soil into watercourses or leached through soil into groundwater

Eutrophication – Many definiton ,one meaning

- Eutrophication is defined as an increase in the rate of supply of organic matter in an ecosystem
- "The term 'eutrophic' means well-nourished; thus, 'eutrophication' refers to natural or artificial addition of nutrients to bodies of water and to the effects of the added nutrients.... When the effects are undesirable, eutrophication may be considered a form of pollution
- Eutrophication or more precisely hypertrophication, is the ecosystem response to the addition of artificial or natural substances, such as nitrates and phosphates, through fertilizers or sewage, to an aquatic system

Fertilizers –> Eutrophicaton ?

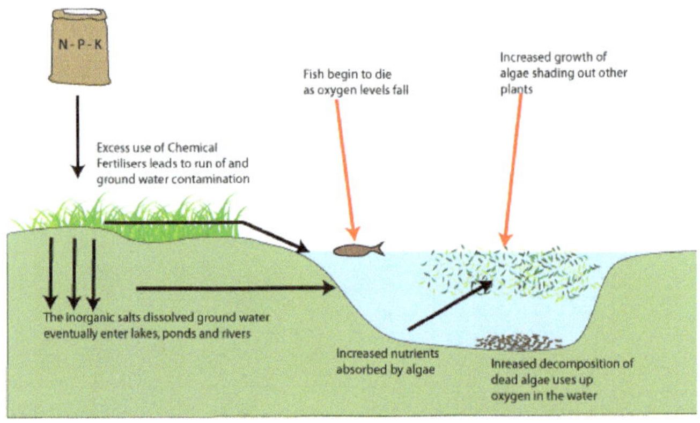

Negative environmental effects include hypoxia, the depletion of oxygen in the water, which induces reductions in specific fish and other animal populations. Other species (such as Nomura's jellyfish in Japanese waters) may experience an increase in population that negatively affects other species.

Impacts of Eutrophication

- In pure state water is odourless and tasteless .However, during the process of eutrophication algae or decaying aquatic vegetation can cause problems of odours

- The world-wide occurrence of toxic algal blooms in fresh, marine and brackish waters creates problems for all life forms. For humans, the health risk of exposure to algal toxins comes above all from drinking water and from the recreational water use.
- Changes in the freshwater ecosystem caused by eutrophication indirectly affect the fish community. Low oxygen conditions can lead to the movement of fish out of affected areas. This leads to reduction of populations of these species, which are not able to tolerate new conditions
- Livestock can also be at risk of cyanotoxins. There are documented many cases of lethal poisonings of livestock after drinking water with high content of toxic cyanobacterial bloom. Every year cattle deaths are associated with cyanobacterial blooms.

Drinking Water

- The main problem associated with high nitrate concentration in drinking water is the development of methaemoglobinaemia. The reactions can take place in stomach or in saliva. Bacteria present there can easily convert nitrate to nitrite. The nitrite combines with haemoglobin in red blood cells and forms methaemoglobin. Methaemoglobin is not able to carry oxygen and it leads to reduce oxygen uptake in the lungs. In pregnant women, haemoglobin is more likely to be converted to methaemoglobin. The people with deficiency of the special enzyme have low ability to convert the methaemoglobin back to haemoglobin.

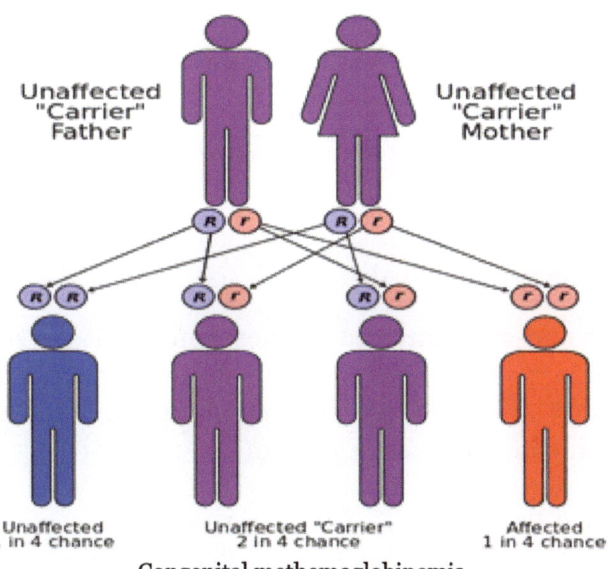

Congenital methemoglobinemia

Blue Baby Syndrome

It is believed to be caused by high nitrate contamination in ground water resulting decreased oxygen carrying capacity of hemoglobin in babies leading to death. Infants under 6 months of age are particularly susceptible to methemoglobinemia caused by nitrates ingested in drinking water dehydration usually caused by gastroenteritis with diarrhea, sepsis, and topical anesthetics containing benzocaine or prilocaine.

- Nitrate and nitrite are able to react with some chemicals in the body and form compounds called N-nitrosamines. These compounds have been proved to cause cancer in animals, which suggests that they may also cause cancer in humans

Soil Pollution

- Soil acidification is the build-up of hydrogen cations, also called protons, reducing the soil such as aluminium sulfate, which reacts in the soil to release protons. Many nitrogen compounds, which are added as fertilizer, acidify soil over the long term because they produce nitrous and nitric acid when oxidized in the process of nitrification.

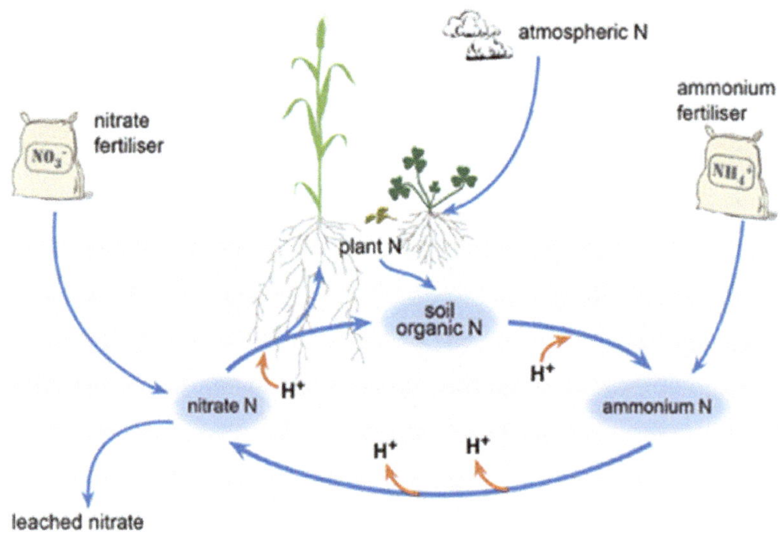

- Contamination with impurities such as fluorides, cadmium and uranium
- The concentration of up to 100 mg/kg of cadmium in phosphate minerals (for example, minerals from Nauru and the Christmas islands increases the contamination of soil with cadmium, for example in New Zealand
- Radioactive Polonium-210 contained in phosphate fertilizers is absorbed by the roots of plants and stored in its tissues; tobacco derived from plants fertilized by rock phosphates contains Polonium-210 which emits alpha radiation estimated to cause about 11,700 lung cancer deaths each year worldwide/Uranium In 2008, CSIC scientists said that nearly half of the catfish showed signs of alarming levels of mercury, and 15-20 per cent exceeded the maximum legal limits. The mercury contamination originates from the Flix swam. The mercury contamination originates from the Flix swamp, where CSIC detected some 700,000 tonnes of toxic mud in 2004
- Toxic persistent organic pollutants ("POPs"), such as Dioxins, polychlorinated dibenzo-p-dioxins (PCDDs), and polychlorinated dibenzofurans (PCDFs) have been detected in agricultural fertilizers and soil amendments. Persistent organic pollutants (POP s) are organic compounds that are resistant to environmental degradation through chemical, biological, and photolytic processes. POP exposure can cause death and illnesses including disruption of the endocrine, reproductive, and immune systems

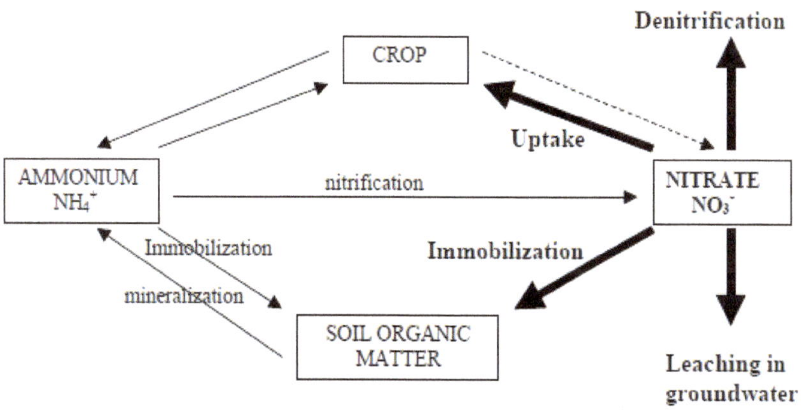

The Effect of Some Common Synthetic Fertilizers on the Soil

Ammonium Sulfate

In the soil, reacts with water to produce sulfuric acid. Sulfuric acid has a pH of less than 1 (2). It is extremely toxic and kills organisms. Hydrogen ions released from the acid replace alkaline elements on the cation exchange sites, depleting the soil of nutrients.

Ammonium Nitrate

In the soil, breaks down into ammonium and nitrate. The ammonium is consumed by plants and fungi, or by denitrifying bacteria which eventually convert it to nitrate. The nitrates are consumed by soil organisms, leached, or converted to nitrogen gas and volatized.

Urea Formaldehyde

The formaldehyde dissolves in water and is leached through the soil. It is a highly toxic substance, killing all soil organisms it comes in contact with. Formaldehyde is used to preserve laboratory specimens

Urea

The urea is consumed by bacteria which convert it to (excrete) anhydrous ammonia (which is a gas) and carbon dioxide (= $2(NH_3) + CO_2$) (9). Anhydrous ammonia is highly toxic and kills organisms. If urea is incorporated into the soil, the ammonia gas reacts with water (H_2O) to produce ammonium hydroxide (NH_4OH), which has a pH of 11.6 (12). It is highly caustic and causes severe burns. This creates a toxic zone in the immediate vicinity of the applied urea that kills seeds, seedlings and soil dwelling organisms.

References

- "Research | AMRF/ORV Alguita Research Projects" Algalita Marine Research Foundation. Macdonald Design. Retrieved 19 May 2009
- Suthar, S; Bishnoi, P; Singh, S; et al. (2009). "Nitrate contamination in groundwater of some rural areas of Rajasthan, India". Journal of Hazardous Materials. 171 (1–3): 189–199. doi:10.1016/j.jhazmat.2009.05.111

- Rand, Gary M.; Petrocelli, Sam R. (1985). Fundamentals of aquatic toxicology: Methods and applications. Washington: Hemisphere Publishing. ISBN 0-89116-382-4
- "About the Organisation for Economic Co-operation and Development (OECD)" Organisation for Economic Co-operation and Development, Retrieved 2012-06-07
- Schnoor J., 1996, Environmental Modeling, Fate and Transport of Pollutants in Water, Air and Soil, Wiley-Interscience,ISBN 978-0-471-12436-8
- ^ATSDR (US Agency for Toxic Substance & Disease Registry) (2008). "Follow-up Health Consultation: Anniston Army Depot." (PDF). Retrieved 18 March 2017
- Singh, B; Singh, Y; Sekhon, GS (1995). "Fertilizer-N use efficiency and nitrate pollution of groundwater in developing countries". Journal of Contaminant Hydrology. 20 (3–4): 167–184. doi:10.1016/0169-7722(95)00067-4
- Lin SD, Lee CC (2001). Water and Wastewater Calculations Manual. McGraw Hill Professional. pp. 13–. ISBN 978-0-07-137195-7
- State Coalition for Remediation of Drycleaners (2011). "A Citizen's Guide to Drycleaner Cleanup" (PDF). Retrieved 20 March 2017
- Smith, M; Cross, K; Paden, M; Laben, P, eds. (2016). Spring - managing groundwater sustainably (PDF). IUCN. ISBN 978-2-8317-1789-0
- "Texas fracking site that spilled 42,000 gallons of fluid into residential area hopes to reopen". RT International. Retrieved 2016-05-07
- Pfunt,H; Houben,G; Himmelsbach,T (2016). "Numerical modeling of fracking fluid migration through fault zones and fractures in the North German Basin". Hydrogeology Journal. 24 (6): 1343–1358. doi:10.1007/s10040-016-1418-7
- Fawell, J; Bailey, K; Chilton, J; Dahi, E (2006). Fluoride in drinking-water (PDF). Geneva: IWA for WHO. ISBN 9241563192
- "Lack of data on fracking spills leaves researchers in the dark on water contamination". StateImpact Pennsylvania. Retrieved 2016-05-09
- Ross, N, ed. (2010). Clearing the waters a focus on water quality solutions. Nairobi, Kenya: UNEP. ISBN 978-92-807-3074-6
- United Nations Environment Programme (UNEP) (2015). "Good Practices for Regulating Wastewater Treatment" (PDF). Retrieved 19 March 2017
- Ahnert, A.; Borowski, C. (2000). "Environmental risk assessment of anthropogenic activity in the deep-sea". Journal of Aquatic Ecosystem Stress and Recovery. 7 (4): 299–315. doi:10.1023/A:1009963912171
- Xu, Y; Usher, B, eds. (2006). Groundwater pollution in Africa (PDF). Taylor & Francis. ISBN 0-415-41167-X
- World Health Organization (WHO) (2006). "Protecting Groundwater for Health - Understanding the drinking-water catchment" (PDF). Retrieved 20 March 2017

4

Soil Formation and Soil Erosion

The major components of soil and its environmental issues are discussed in this section. There are various elements that lead to the formation of soil starting from climate, parent material, organisms, time and topography. Soil is composed of minerals, plant roots, organic materials, and pore spacing. The chapter also explores the concept of soil erosion and other environmental issues related to soil.

Soil and its Formation

Pedosphere as Environmental Interface

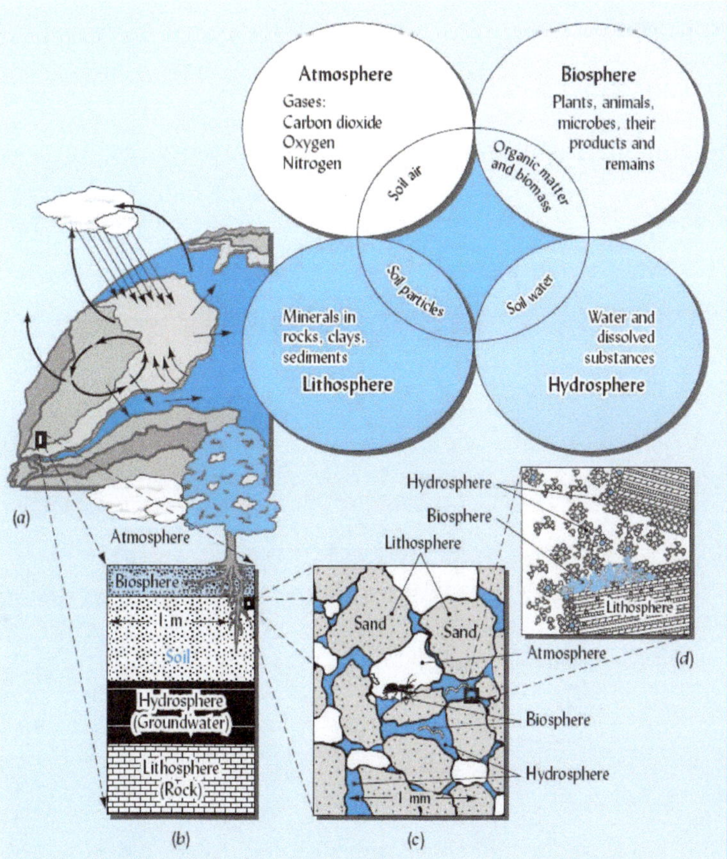

The pedosphere—interface of the worlds of rock (the lithosphere), air (the atmosphere), water (the hydrosphere), and life (the biosphere)—can be understood at many different scales. At the kilometer scale (a), soil participates in global cycles and the life of terrestrial ecosystems. At the meter scale (b), soil forms a transition zone between the hard rock below and the atmosphere above—a zone through which surface water and groundwater flow and in which plants and other living organisms thrive. At the millimeter scale (c), mineral particles form the skeleton of the soil that defines pore spaces, some filled with air and some with water, in which tiny creatures lead their lives. Finally, at the micro- and nanometer scales (d), soil minerals (lithosphere) provide charges, reactive surfaces that adsorb water and cations dissolved in water (hydrosphere), gases (atmosphere), and bacteria and complex humus macromolecules (biosphere). (Diagram courtesy of R. Weil)

Soils so deeply affect every other part of the ecosystem that they are often called the "great integrator." M. J. Singer argued that soils are ecosystems by themselves.

Soils have six key roles to play:

- Support plant growth;
- Reserve and purify water;
- Function as nature's recycling systems;
- Regulate the atmosphere;
- Provide habitats for a myriad of living organisms;
- Supply engineering materials and building foundations

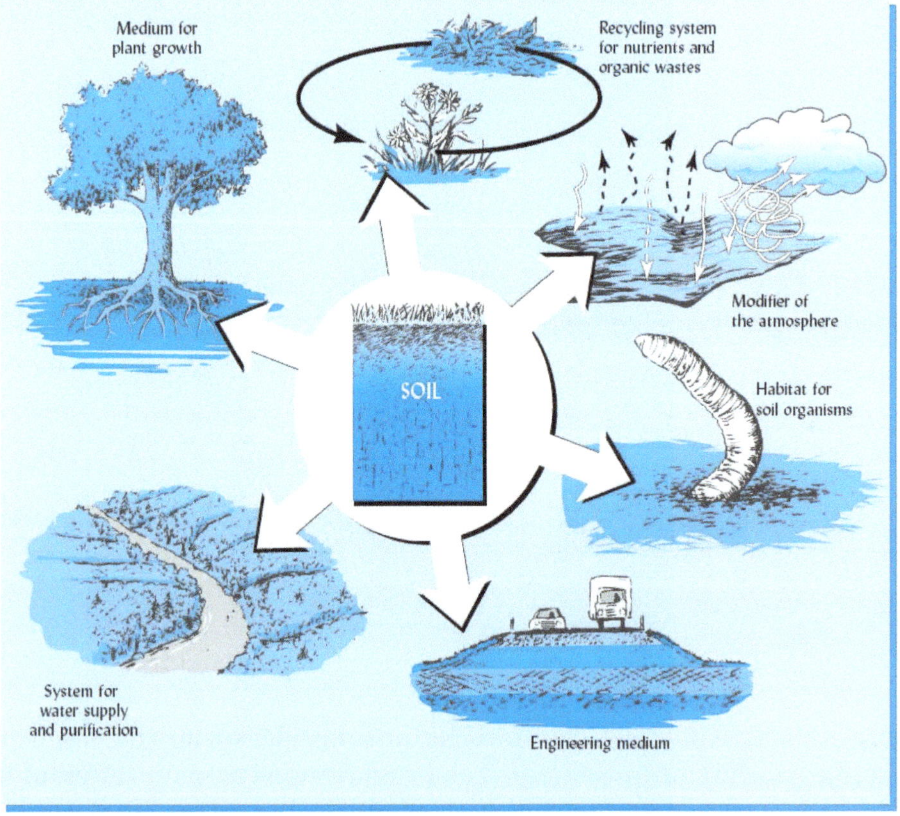

The main functions of soil can be grouped into six crucial ecological roles.

Soils as Media for Plant Growth

ELEMENTS ESSENTIAL FOR PLANT GROWTH AND THEIR SOURCES[a]

The chemical forms most commonly taken in by plants are shown in parentheses, with the chemical symbol for the element in bold type.

Macronutrients: Used in relatively large amounts (>0.1% of dry plant tissue)		Micronutrients: Used in relatively small amounts (<0.1% of dry plant tissue)
Mostly from air and water	Mostly from soil solids	From soil solids
Carbon (CO_2) Hydrogen (H_2O) Oxygen (O_2, H_2O)	*Cations:* Calcium ($\mathbf{Ca}^{2+}$) Magnesium ($\mathbf{Mg}^{2+}$) Nitrogen ($\mathbf{N}H_4^+$) Potassium ($\mathbf{K}^+$)	*Cations:* Copper ($\mathbf{Cu}^{2+}$) Iron ($\mathbf{Fe}^{2+}$) Manganese ($\mathbf{Mn}^{2+}$) Nickel ($\mathbf{Ni}^{2+}$) Zinc ($\mathbf{Zn}^{2+}$)
	Anions: Nitrogen ($\mathbf{N}O_3^-$) Phosphorus ($H_2\mathbf{P}O_4^-$, $H\mathbf{P}O_4^{2-}$) Sulfur ($\mathbf{S}O_4^{2-}$)	*Anions:* Boron ($H_3\mathbf{B}O_3$, $H_4\mathbf{B}O_4^-$) Chlorine ($\mathbf{Cl}^-$) Molybdenum ($\mathbf{Mo}O_4^{2-}$)

[a] Many other elements are taken up from soils by plants but are not *essential* for plant growth (see Epstein and Bloom, 2005).

Soil as a Natural Body

The soil is often said to cover the land, and is a collection of individually different soil bodies.

A soil is an individual tree to a forest, a three-dimensional natural body.

Soils are often used, in general, to refer to all individual natural bodies as a whole.

A regolith is a layer of unconsolidated materials overlaying the hard, unweathered, bedrock.

- Some History of Soil Science:

  (1) The "willow" experiment by Jan Baptiste Van Helmont (1577-1644) in Holland. He put 5-lb willow tree in 200 lb of soil. The willow tree received only rainwater for 5 years. At the end of 5 years, the tree weighed 169 lb 3oz, the soil weighed 57 g less than 200 lb. So he reasoned that water was the "principal of vegetation."

  (2) Justus Liebig's law of the minimum (mid 1800) (Also Carl Sprengel).

  (3) The Hatch Act in the US in 1862.

- The Composition

  Soils are composed of four main components:

  (1) Minerals of different sizes (clay, silt, & sand);

  (2) Organic materials from the remains of dead plants and animals;

  (3) Open space that can be filled with water and air (pore space);

  (4) Plant roots and other living organisms.

- Soil formation and properties:

  (1) parent material --The material from which the soil is formed. Soil parent material could be bedrock, organic material, or a deposit from water, wind, glaciers, volcanoes, or material moving down a slope.

  (2) climate -- Heat, rain, ice, snow, wind, sunshine, and other environmental forces break down the parent material and affect how fast or slow soil processes go.

  (3) organisms -- All plants and animals living in or on the soil (including micro- organisms and humans!).

  (4) topography -- The location of a soil on a landscape can affect how the climatic processes impact it. Soils at the bottom of a hill will get more water than soils on the slopes, and soils on the slopes that directly face the sun will be drier than soils on slopes that do not.

  (5) time -- All of the above factors assert themselves over time, often hundreds or thousands of years.

- Soil Profiles

The face of a soil, or the way it looks if you cut a section of it out of the ground, is called a soil profile. The profile can tell you about the geology and climate history of the landscape over thousands of years, the archeological history of how humans used the soil, what the soil's properties are today, and the best way to use the soil.

Every soil profile is made up of layers called soil horizons. Soil horizons can be as thin as a few millimeters or thicker than a meter.

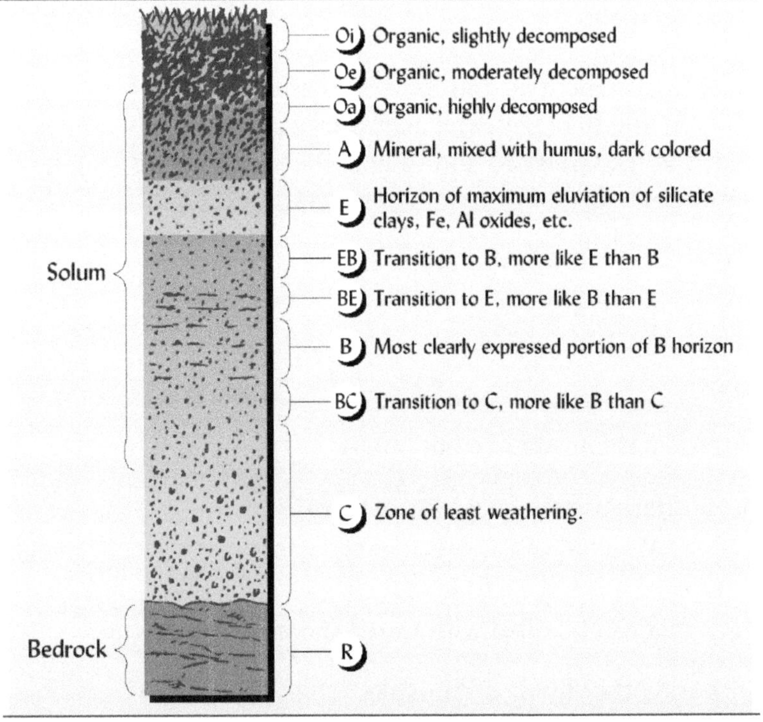

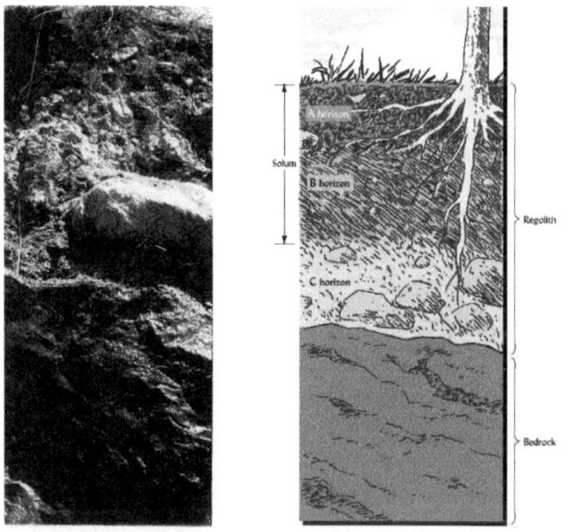

Relative positions of the regolith, its soil, and the underlying bedrock. Note that the soil is a part of the regolith, and that the A and B horizons are part of the *solum* (from the Latin word *solum*, which means soil or land). The C horizon is the part of the regolith that underlies the solum, but may be slowly changing into soil in its upper parts. Sometimes the regolith is so thin that it has been

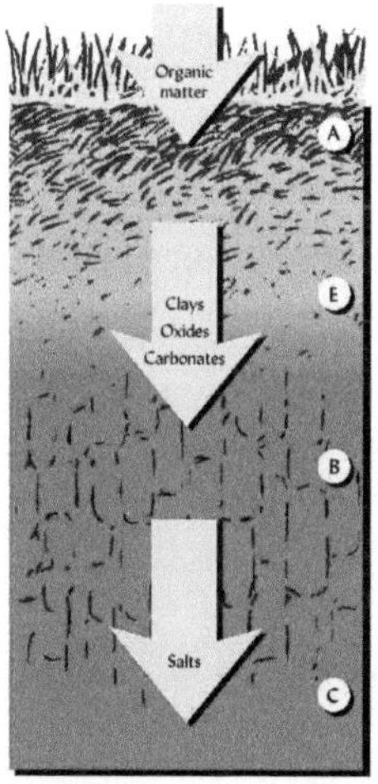

Horizons begin to differentiate as materials are added to the upper part of the profile and other materials are translocated to deeper zones. Under certain conditions, usually associated with forest vegetation and high rainfall, a leached E horizon forms between organic-matter-rich A and the B horizons. If sufficient rainfall occurs, soluble salts will be carried below the soil profile, perhaps all the way to the groundwater.

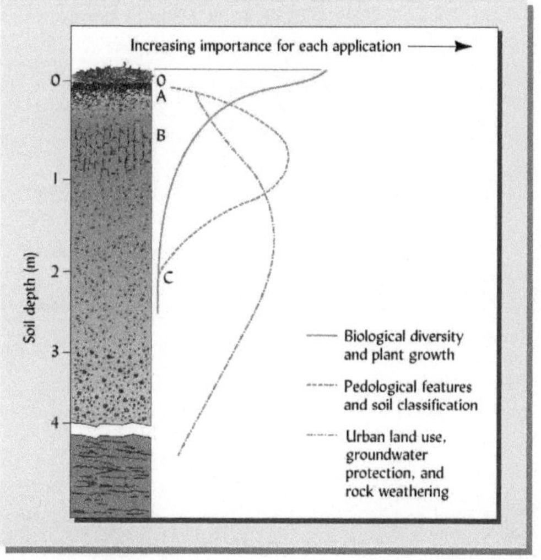

Information important to different soil functions and applications is most likely to be obtained by studying different layers of the soil profile.

Minnesota Spotoso

Australia Oxiso

Germany Spotoso

Soil Characterization or Properties

Soil structure refers to the natural shape of groups of soil particles or aggregates (peds) in the soil. The structure affects how big the spaces will be in the soil through which roots, air, and water may move.

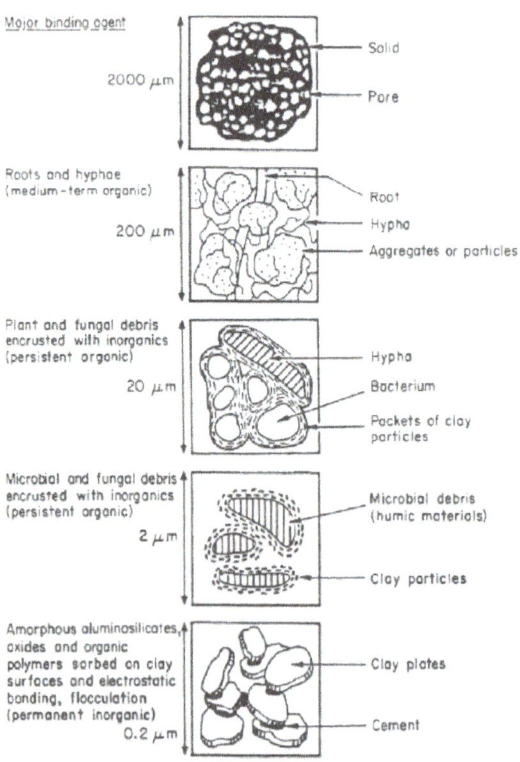

Soil Texture is how the soil feels and is determined by the amount of sand (2.00- 0.05 mm), silt (0.05-0.002 mm), and clay (<0.002) particles in the soil, each of which is a different size. The actual amount of sand, silt, and clay size particles in a soil sample is also called the particle size distribution.

The relative distribution of sand, silt, and clay particles determines soil texture.

General Properties of the Three Major Size Classes of Inorganic Soil Particles

	Property	Sand	Silt	Clay
1.	Range of particle diameters in mm	2.0–0.05	0.05–0.002	Smaller than 0.002
2.	Means of observation	Naked eye	Microscope	Electron microscope
3.	Dominant minerals	Primary	Primary and secondary	Secondary
4.	Attraction of particles for each other	Low	Medium	High
5.	Attraction of particles for water	Low	Medium	High
6.	Ability to hold chemicals and nutrients in plant-available form	Very low	Low	High
7.	Consistency when wet	Loose, gritty	Smooth	Sticky, malleable
8.	Consistency when dry	Very loose, gritty	Powdery, some clods	Hard clods

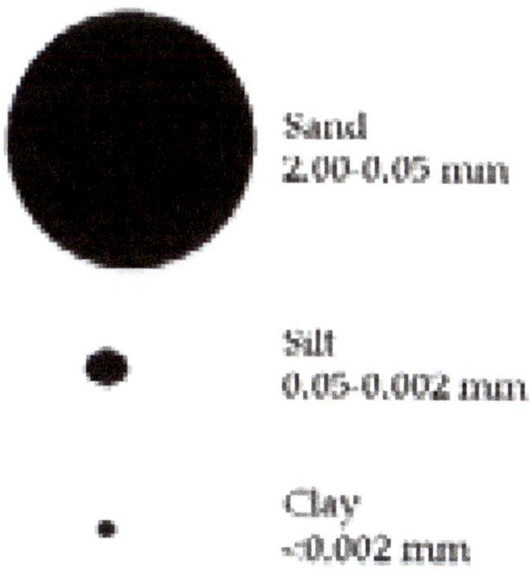

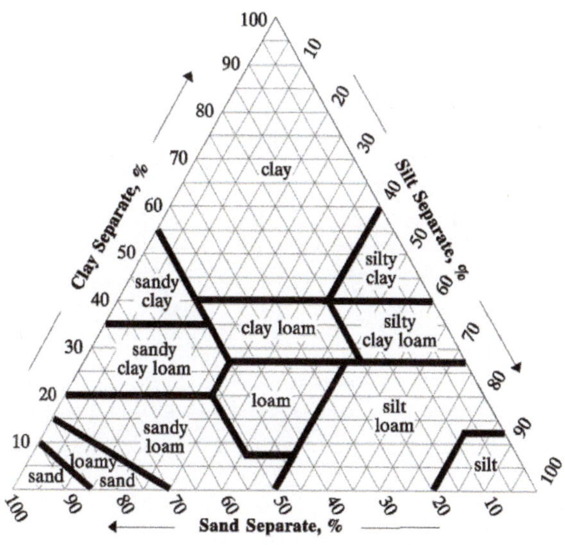

Bulk Density:

Soil bulk density is a measurement of how tightly packed or dense the soil is. It is determined by measuring the weight of dry soil in a unit of volume (g/cm³).

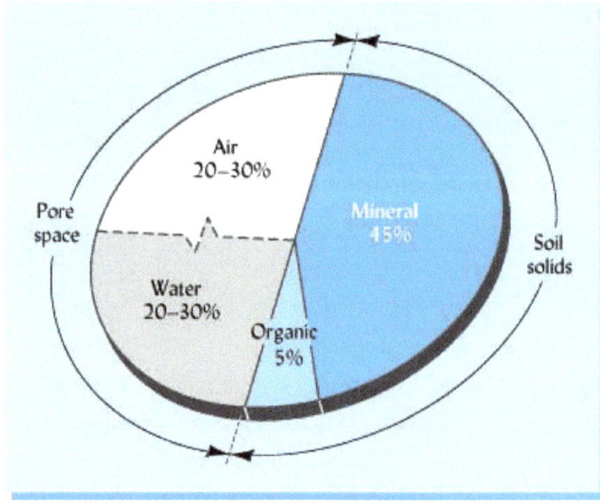

pH: molar $[H_3O^+] = 10^{-pH}$; $H_2O + H_2O \Leftrightarrow H_3O^+ + OH^-$

Soil pH influences what can grow in the soil and is the product of the kind of parent material, the chemical nature of the rain and other water entering the soil, land management practices, and the activities of organisms (plants, animals, fungi, protists, and monera) living in the soil.

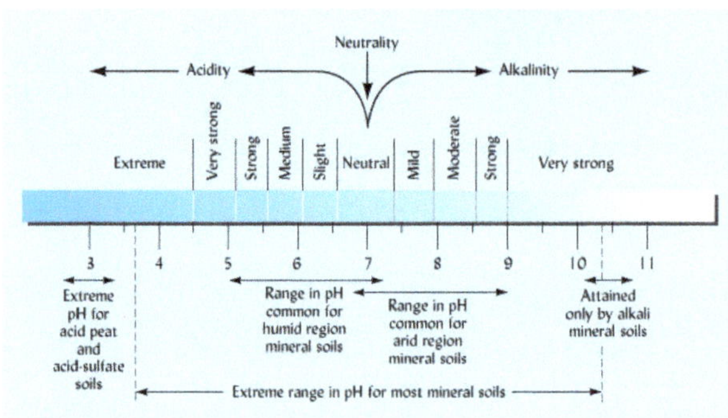

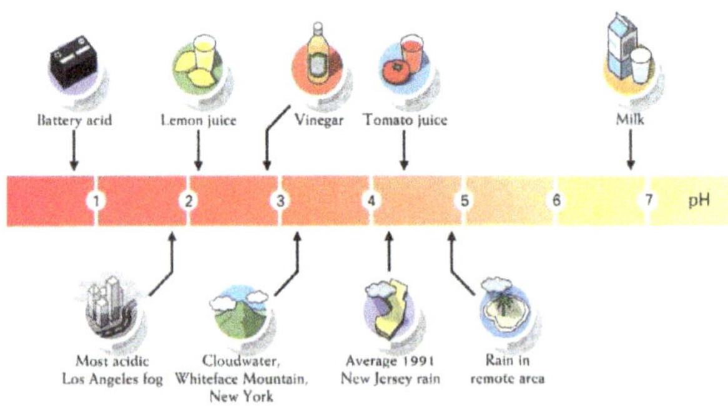

The pH values in atmospheric water of various types, compared with the pH values for several common liquids.

Soil Moisture

Soil moisture is a measurement of water content, or water availability. Soil water content typically ranges between 0.05 and 0.40 g/g (θ_m gravimetric).

θ_m = gravimetric water content = mass of water per unit mass of dry soil (g/g; kg/kg etc.);

θ_v = volumetric water content = volume of water per unit volume of soil; (cm^3/cm^3; m^3/m^3).

The fertility of a soil is determined by how many kinds and how much nutrients it has stored and is able to supply. Nitrogen (N) mostly in the form of nitrate (NO_3-) or ammonium (NH_4+), phosphorus (P in the form of H_2PO_4- or HPO_4-), and potassium (K as K^+) are three often mentioned soil nutrients important for the growth of plants.

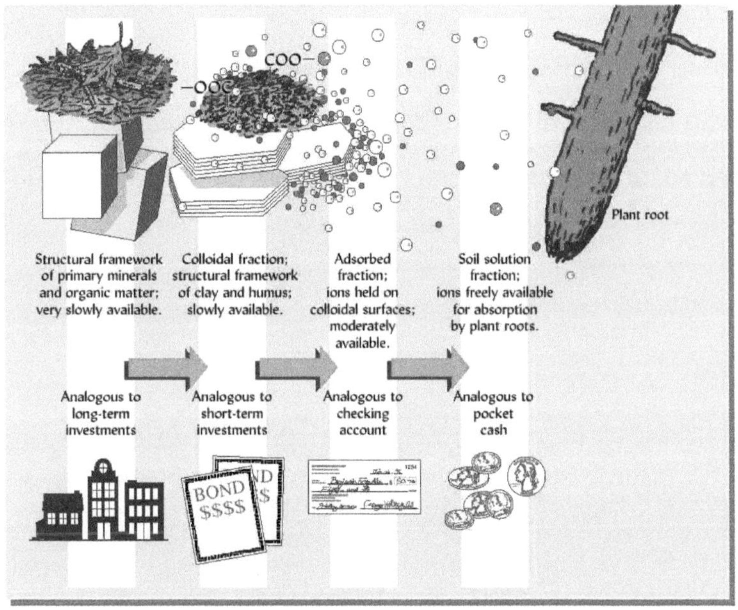

Nutrient elements exist in soils in various forms characterized by different accessibility to plant roots. The bulk of the nutrients is locked up in the structural framework of primary minerals, organic matter, clay, and humus. A smaller proportion of each nutrient is adsorbed in a swarm of ions near the surfaces of soil colloids (clay and organic matter). From the swarm of adsorbed ions, a still smaller amount is released into the bulk soil solution, where uptake by plant roots can take place.

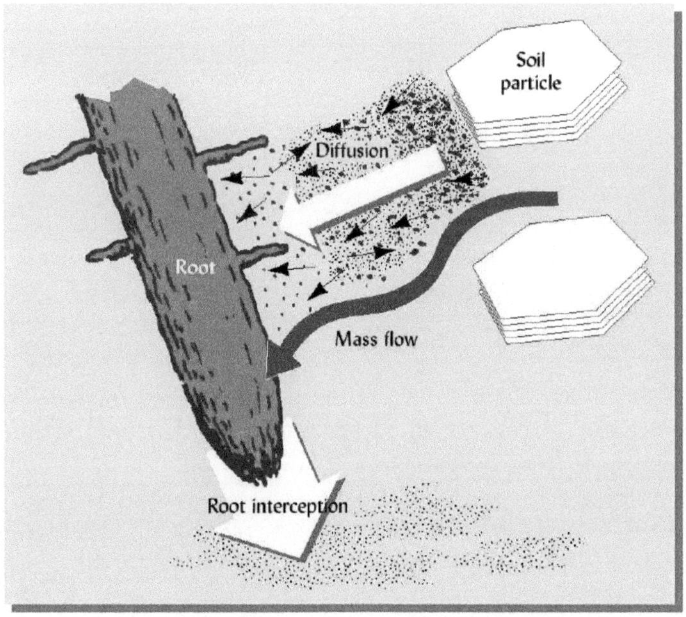

Three principal mechanisms by which nutrient ions dissolved in the soil solution come into contact with plant roots. All three mechanisms may operate simultaneously, but one mechanism or another may be most important for a particular nutrient. For example, in the case of calcium, which is generally plentiful in the soil solution, mass flow alone can usually bring sufficient amounts to the root surface. However, in the case of phosphorus, diffusion is needed to supplement mass flow because the soil solution is very low in this element in comparison to the amounts needed by plants.

Importance of Water Concentration

Why is water important for soil systems?

1. Properties of water

Molecular polarity and hydrogen bonding

Cohesion: the attraction of water molecules between each other primarily due to hydrogen bonding.

Adhesion (also called adsorption): the attraction of water molecules for solid surfaces.

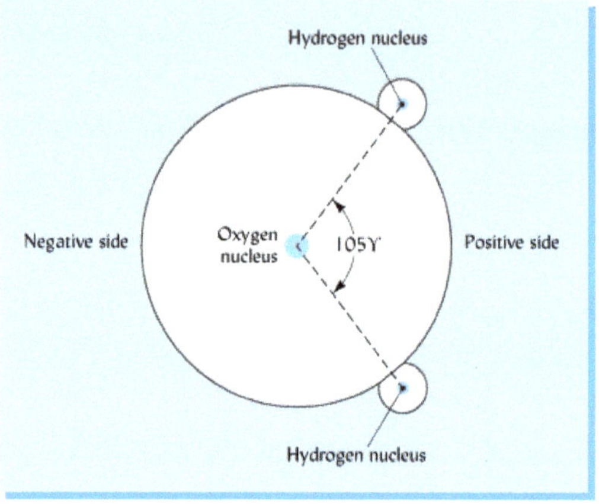

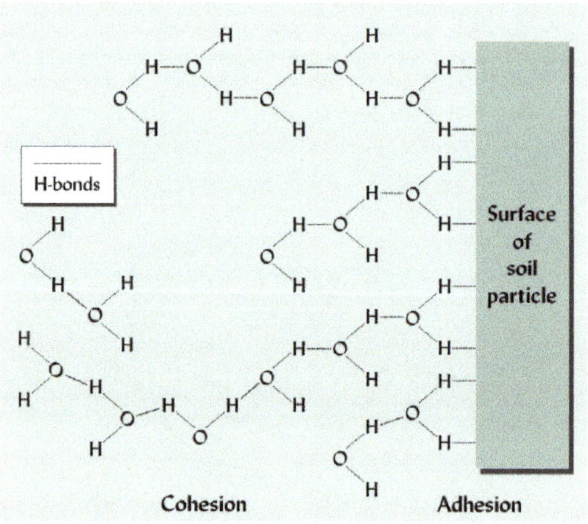

2. Capillarity

Surface tension: the greater attraction of water molecules for each other than the air above at liquid-air interfaces primarily due to cohesion.

- Adhesion and surface tension together cause the phenomenon called capillarity--the movement of water up a wick made of hydrophilic solid materials.
- Capillary movement takes place in any direction.
- The height of capillary rise in a tube is directly proportional to the liquid's surface tension and adhesion to the solid surface, but inversely proportional to the tube radius and the density of the liquid.

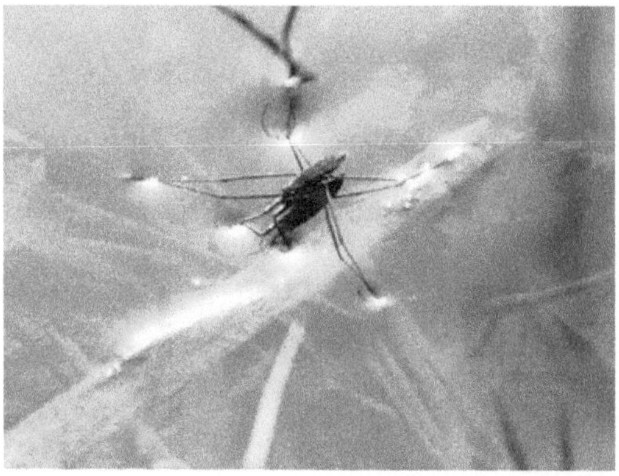

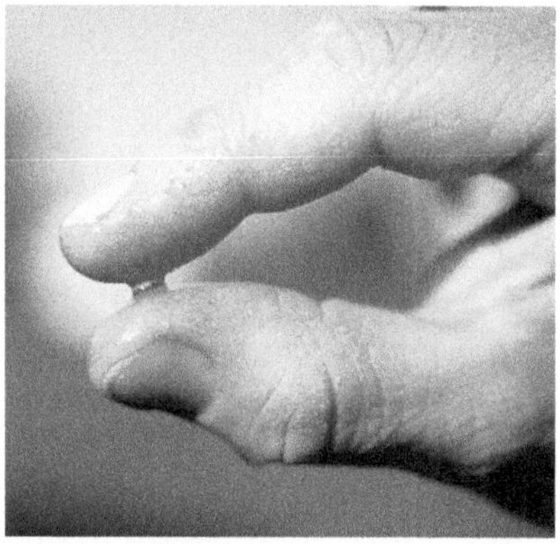

Everyday evidences of water's surface tension (left) as insects walk on water and do not sink, and of forces of cohesion and adhesion (right) as a drop of water is held between the fingers. (Photos courtesy of R. Weil)

As this field irrigation scene in Arizona shows (Top), water has moved up by capillarity from the irrigation furrow to-ward the top of the ridge. The photo on the Bottom illustrates some horizontal movement to both sides and away from the irrigation water.

3. Soil water content and water potential

Gravimetric water content: the weight of water in a soil sample per unit weight of dry soil, such as 0.25 kg water per kg of dry soil.

Volumetric water content: the volume of water in a soil sample per unit of total soil volume.

- When considering soil water availability and movement, a general energy expression of soil water status is often used, i.e., the tendency of natural movement for all substances is always from higher energy stage to a lower energy stage. So does for water. Higher energy stage of water often means higher availability for biological use.

When soil water status is expressed in energy forms and compared to pure water under standard temperature and pressure, the difference in energy level between this pure water and that of soil water is called soil water potential.

Yt = Ym + Ys + Yp + Yg

Ym : Matric potential

Ys : Solute potential or osmotic potential

Yp : Pressure potential or submergence

Yg : Gravitational or elevation gradient

Please refer, for information, about methods for measuring soil water.

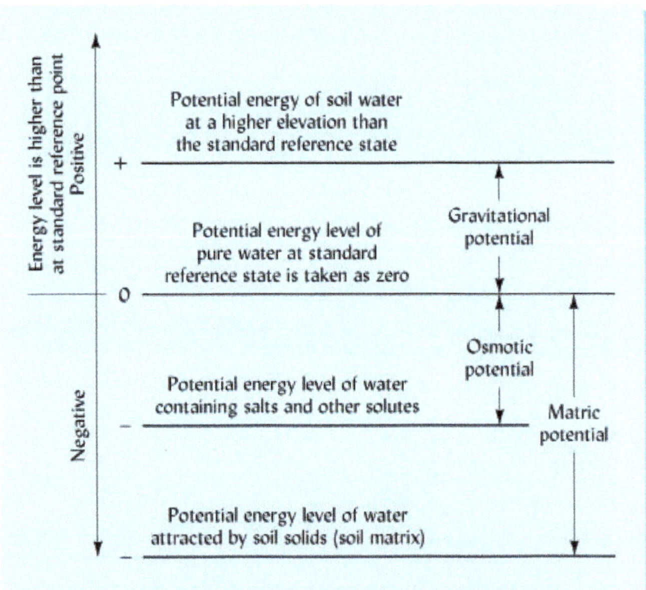

Relationship between the potential energy of pure water at a standard reference state (pressure, temperature, and elevation) and that of soil water. If the soil water contains salts and other solutes, the mutual attraction between water molecules and these chemicals reduces the potential energy of the water, the degree of the reduction being termed osmotic potential. Similarly, the mutual attraction between soil solids (soil matrix) and soil water molecules also reduces the water's potential energy. In this case the reduction is called matric potential. Since both of these interactions reduce the water's potential energy level compared to that of pure water, the changes in energy level (osmotic potential and matric potential) are both considered to be negative. In contrast, differences in energy due to gravity (gravitational potential) are always positive. This is because the reference elevation of the pure water is purposely designated at a site in the soil profile below that of the soil water. A plant root attempting to remove water from a moist soil would have to overcome all three forces simultaneously.

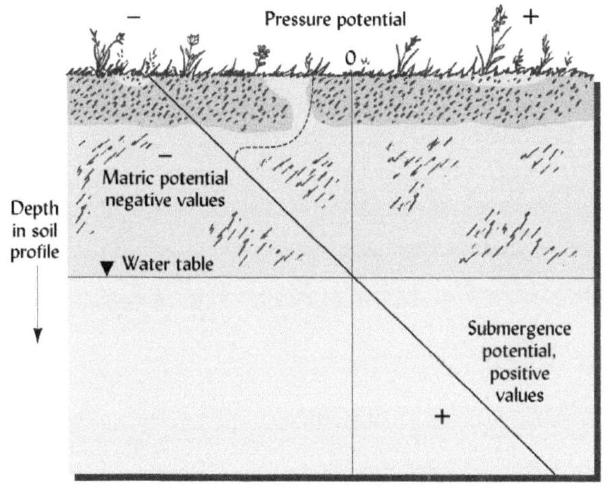

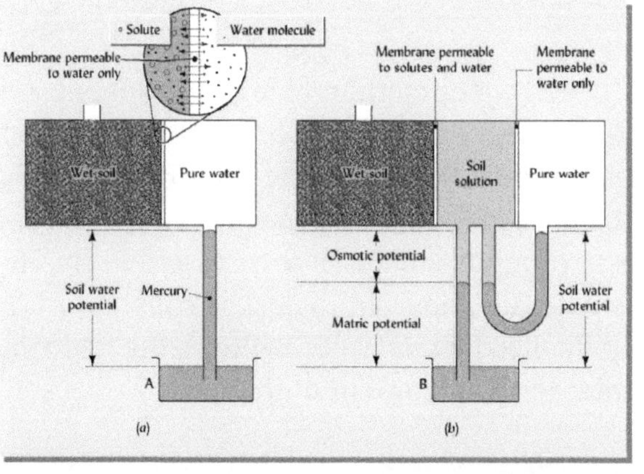

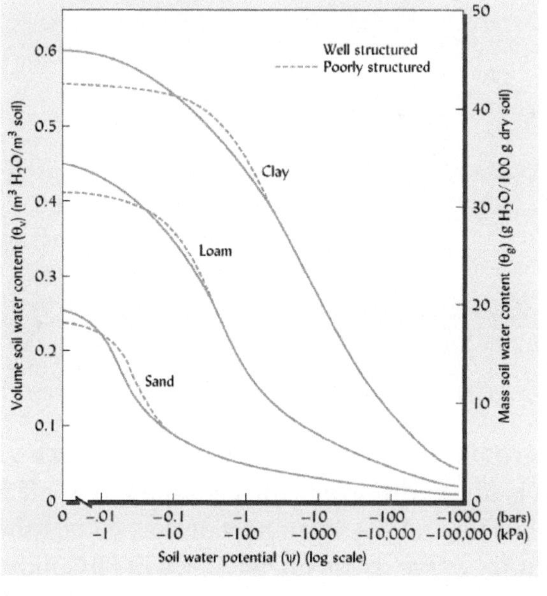

Some Methods of Measuring Soil Water

More than one method may be needed to cover the entire range of soil moisture conditions.

Method	Measures soil water: Content	Measures soil water: Potential	Useful range, kPa	Used mainly in: Field	Used mainly in: Lab	Comments
1. Gravimetric	X		0 to < −10,000		X	Destructive sampling; slow (1 to 2 days) unless microwave used. The standard for calibration.
2. Neutron scattering	X		0 to < −1500	X		Radiation permit needed; expensive equipment; not good in high-organic-matter soils; requires access tube.
3. Time domain reflectometry (TDR)	X		0 to < −10,000	X	X	Can be automated; accurate to ±1 to 2% volumetric water content; very sandy or salty soils need calibration; requires wave guides; expensive instrument.
4. Capacitance sensors	X		0 to < −1500	X	X	Can be automated; accurate to ±2 to 4% volumetric water content; sands or salty soils need calibration; simple, inexpensive sensors and recording instruments.
5. Resistance blocks		X	− 90 to < −1500	X	X	Can be automated; not sensitive near optimum plant water contents; may need calibration.
6. Tensiometer		X	0 to −85	X	X	Can be automated; accurate to ±0.1 to 1 kPa; limited range; inexpensive; needs periodic servicing to add water.
7. Thermocouple psychrometer		X	50 to < −10,000	X	X	Moderately expensive; wide range; accurate only to ±50 kPa.
8. Pressure membrane apparatus		X	50 to < −10,000		X	Used with gravimetric method to construct drier part of water characteristic curve.
9. Tension table		X	0 to −50		X	Used with gravimetric method to construct wetter part of water characteristic curve.

Instrumental measurement of soil water content using time domain reflectometry (TDR). The instrument sends a pulse of electromagnetic energy down the two parallel metal rods of a waveguide that the soil scientist is pushing into the soil (inside photo). The TDR instrument makes precise picosecond measurements of the speed at which the pulse travels down the rods, a speed influenced by the nature of the surrounding soil. Microprocessors in the instrument analyze the wave patterns generated and calculate the apparent dielectric constant of the soil. Since the dielectric constant of a soil is mainly influenced by its water content, the instrument can accurately convert its measurements into volumetric water content of the soil.

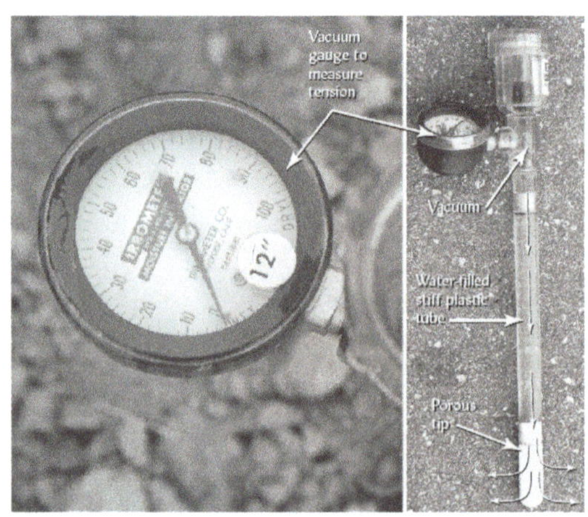

Tensiometer is used to determine water potential in the field. The side view (right) shows the entire instrument. The tube is filled with water through the screw-off top. Once the instrument is tightly sealed, the white porous tip and the lower part of the plastic tube is inserted into a snug-fitting hole in the soil. The vacuum gauge (close up, left) will directly indicate the tension or negative potential generated as the soil draws the water out (curved arrows) through the porous tip. Note the scale goes up to only 100 centibars (= 100 kPa) tension at the driest.

- Water movement in soils

(1) Saturated flow (Soil texture, structure, organic matter, hardpan, water content, temperature, etc.) :

Darcy's Law: $Q / t = (A\ Ksat\ \Delta\psi) / L$

A = cross sectional area of the column; K_{sat} = saturated hydraulic conductivity

$\Delta\psi/L$ = water potential gradient, or hydraulic gradient

For typical K_{sat} values of some soil types. But please note the preferential flow, and the 104 power relationship with the pore radius changes.

(2) Unsaturated flow (Driven by matric potential gradient, or the thickness of water films on particles, <-10 kPa)

(3) Infiltration: the process by which water enters the soil pore space.

(4) Percolation: the process of water moving downward into the profile after infiltration, which involves both saturated flow and unsaturated flow.

(5) Water vapor movement:

- *Moist →Dry; Cool →Warm*

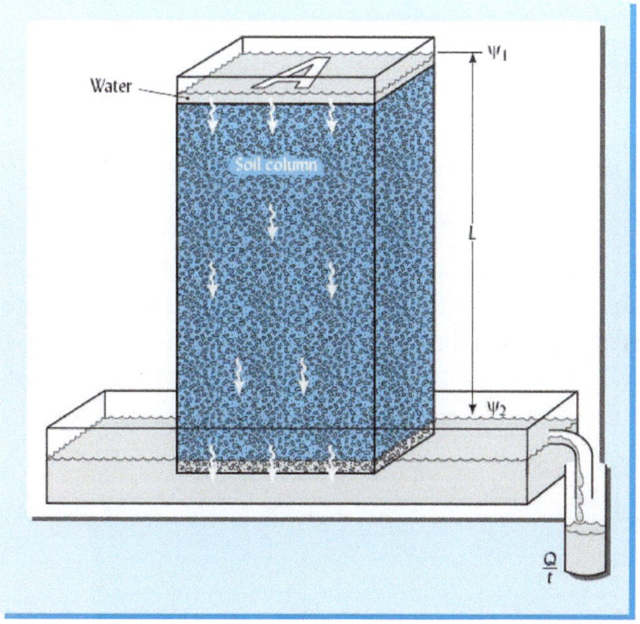

Saturated flow (percolation) in a column of soil with cross-sectional area A, cm². All soil pores are filled with water. At lower right, water is shown running off into a container to indicate that water is actually moving down the column. The force driving the water through the soil is the water potential gradient, $\psi_1 - \psi_2/L$, where both water potentials and length are expressed in cm.

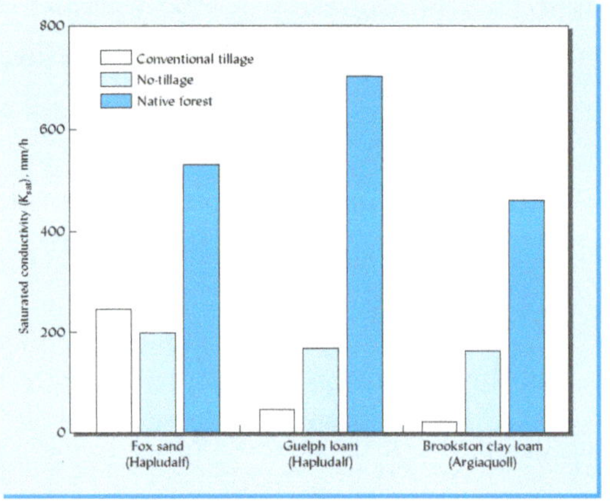

The effect of land management and soil texture on saturated conductivity (Ksat) of three soils in Canada. Soils under native woodlots had higher Ksat values, apparently due to higher organic matter contents and to preferential flow channels provided by decayed roots and burrowing animals. Tillage practices had little effect on conductivity in sand, but in loam and clay loam soils, conductivity was higher where no-tillage systems had been used, suggesting that no-till had increased the proportion of larger, water-conducting pores.

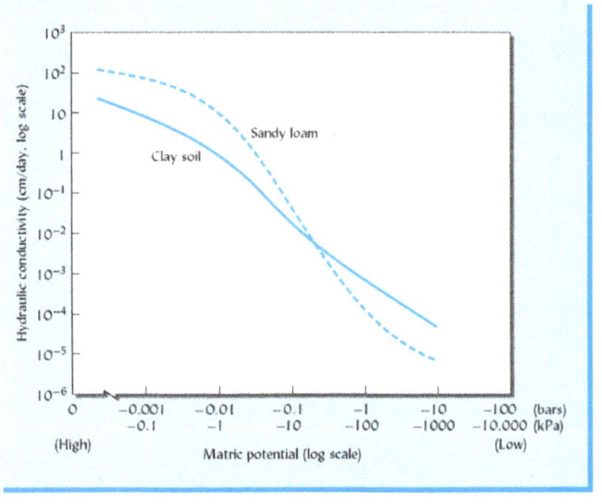

Generalized relationship between matric potential and hydraulic conductivity for a sandy soil and a clay soil (note log scales). Saturated flow takes place at or near zero potential, while much of the unsaturated flow occurs at a potential of -0.1 bar (-10 kPa) or below.

SOME APPROXIMATE VALUES OF SATURATED HYDRAULIC CONDUCTIVITY (IN VARIOUS UNITS) AND INTERPRETATIONS FOR SOIL USES

K_{sat}, cm/s	K_{sat}, cm/h	K_{sat}, in./h	Comments
1×10^{-2}	36	14	Typical of beach sand.
5×10^{-3}	18	7	Typical of very sandy soil, too rapid to effectively filter pollutants in wastewater.
5×10^{-4}	1.8	0.7	Typical of moderately permeable soils, K_{sat} between 1.0 and 15 cm/h considered suitable for most agricultural, recreational, and urban uses calling for good drainage.
5×10^{-5}	0.18	0.07	Typical of fine-textured, compacted, or poorly structured soils. Too slow for proper operation of septic tank drain fields, most types of irrigation, and many recreational uses such as playgrounds.
$<1 \times 10^{-8}$	$<3.6 \times 10^{-5}$	$<1.4 \times 10^{-5}$	Extremely slow; typical of compacted clay. K_{sat} of 10^{-5} to 10^{-8} cm/h may be required where nearly impermeable material is needed, as for wastewater lagoon lining or landfill cover material.

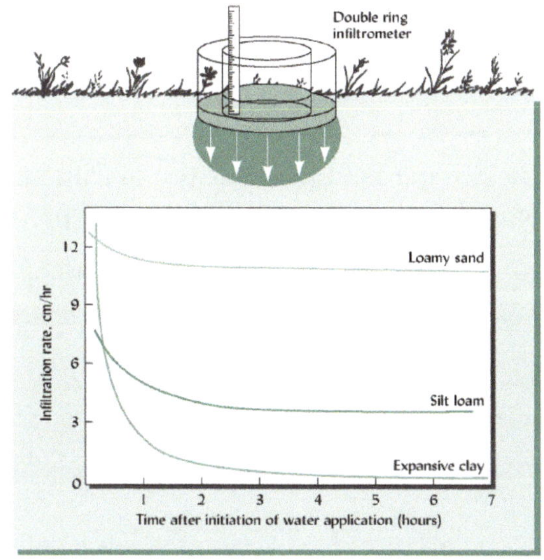

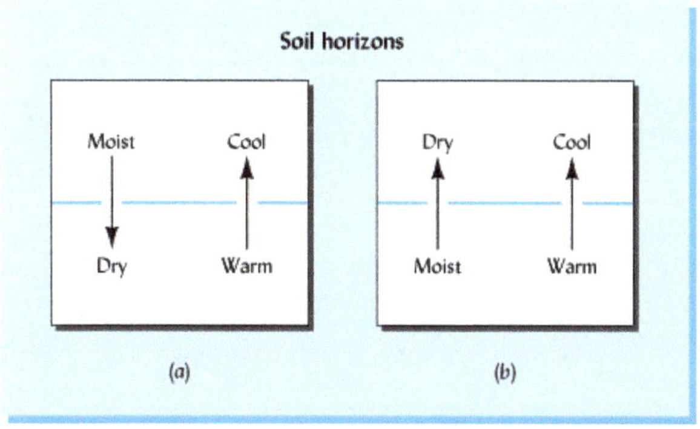

Vapor movement tendencies that may be expected between soil horizons differing in temperature and moisture. In (a) the tendencies more or less negate each other, but in (b) they are coordinated and considerable vapor transfer might be possible if the liquid water in the soil capillaries does not interfere.

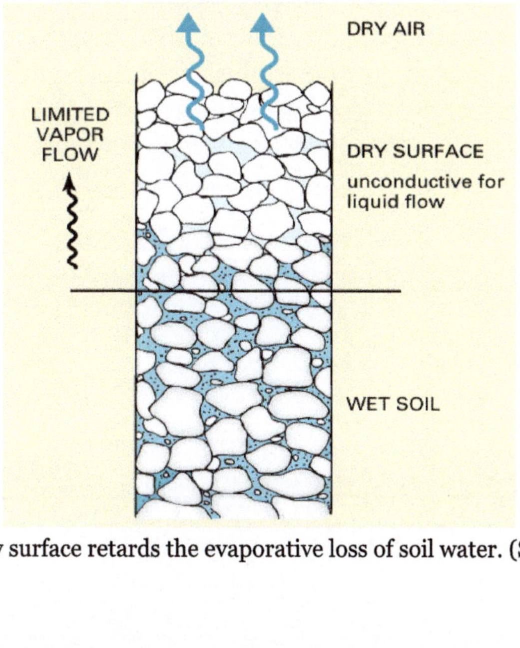

Low conductivity in dry surface retards the evaporative loss of soil water. (Singer & Munns, 2006)

- Qualitative description of soil wetness

(1) Maximum retentive capacity: soil water content when all pores are filled with water, or totally saturated.

(2) Field capacity (or field water holding capacity): maximum soil water content after the drain of macropore water due to gravity gradient (~ -33 Kpa).

(3) Permanent wilting point (or, wilting coefficient, wilting percentage): the soil water potential (content) at or below which plant roots cannot absorb water any more (~ -1500 Kpa).

(4) Hygroscopic coefficient: the soil water content of air-dried soil after reaching equilibrium with water vapor in the air of 98% relative humidity. Clay colloids are the main agents for holding this water.

(5) Factors influencing thresholds above: Texture, SOM, bulk density, and salinity

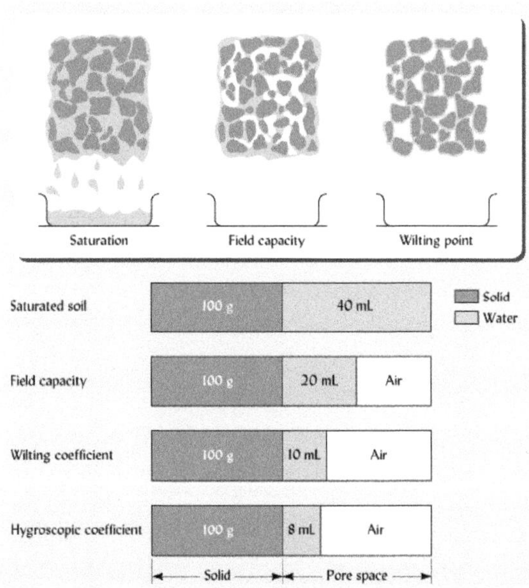

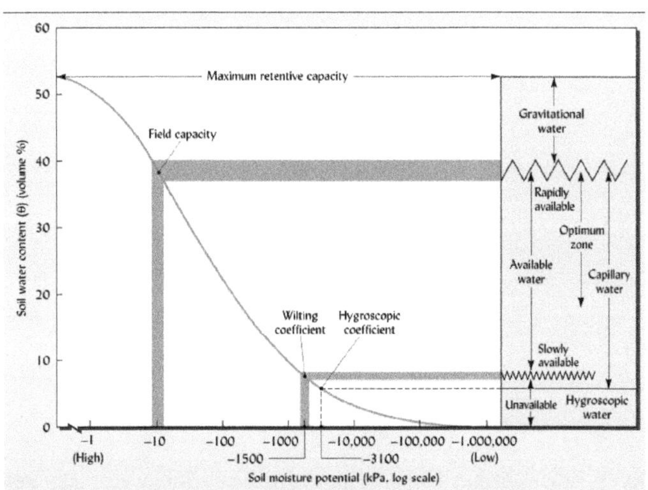

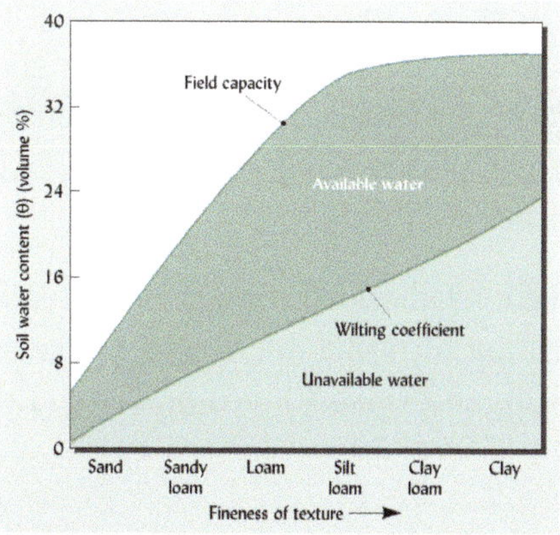

6. Plant water uptake

About 90% by Passive Absorption (or capillary action)Water loss through Evapotranspiration. The maximum ET loss that would occur under soil water content near field capacity is referred to as Potential ET (ETp), which is usually below Pan Evaporation (50-90% of Pan Evaporation)

Water Use Efficiency (WUE): the amount of water required to produce a unit weight of dry plant material, such as the Transpiration Ratio (water/dry matter, ranges from 200:1 to1000:1).

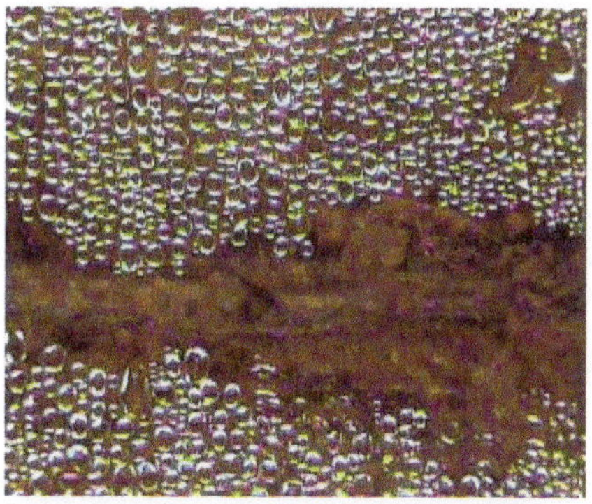

A minirhizotron image of soil water removal by a pine root

Soil Erosion

Soil erosion is the displacement of upper layer of soil, one form of soil degradation. The erosion of soil is a naturally occurring process on all land. The agents of soil erosion are water and wind, each contributing a significant amount of soil loss each year. Soil erosion may be a slow process that continues relatively unnoticed, or it may occur at an alarming rate causing a serious loss of topsoil. The loss of soil from farmland may be reflected in reduced crop production potential, lower surface water quality and damaged drainage networks.

While erosion is a natural process, human activities have increased by 10–40 times the rate at which erosion is occurring globally. Excessive (or accelerated) erosion causes both "on-site" and "off-site" problems. On-site impacts include decreases in agricultural productivity and (on natural landscapes) ecological collapse, both because of loss of the nutrient-rich upper soil layers. In some cases, the eventual end result is desertification. Off-site effects include sedimentation of waterways and eutrophication of water bodies, as well as sediment-related damage to roads and houses. Water and wind erosion are the two primary causes of land degradation; combined, they are responsible for about 84% of the global extent of degraded land, making excessive erosion one of the most significant environmental problems worldwide.

Intensive agriculture, deforestation, roads, anthropogenic climate change and urban sprawl are amongst the most significant human activities in regard to their effect on stimulating erosion. However, there are many prevention and remediation practices that can curtail or limit erosion of vulnerable soils.

Physical Processes

Rainfall and Surface Runoff

Soil and water being splashed by the impact of a single raindrop.

Rainfall, and the surface runoff which may result from rainfall, produces four main types of soil erosion: *splash erosion*, *sheet erosion*, *rill erosion*, and *gully erosion*. Splash erosion is generally seen as the first and least severe stage in the soil erosion process, which is followed by sheet erosion, then rill erosion and finally gully erosion (the most severe of the four).

In *splash erosion*, the impact of a falling raindrop creates a small crater in the soil, ejecting soil particles. The distance these soil particles travel can be as much as 0.6 m (two feet) vertically and 1.5 m (five feet) horizontally on level ground.

If the soil is saturated, or if the rainfall rate is greater than the rate at which water can infiltrate into the soil, surface runoff occurs. If the runoff has sufficient flow energy, it will transport loosened soil particles (sediment) down the slope. *Sheet erosion* is the transport of loosened soil particles by overland flow.

Rill erosion refers to the development of small, ephemeral concentrated flow paths which function as both sediment source and sediment delivery systems for erosion on hillslopes. Generally, where water erosion rates on disturbed upland areas are greatest, rills are active. Flow depths in rills are typically of the order of a few centimeters (about an inch) or less and along-channel slopes may be quite steep. This means that rills exhibit hydraulic physics very different from water flowing through the deeper, wider channels of streams and rivers.

A spoil tip covered in rills and gullies due to erosion processes caused by rainfall: Rummu, Estonia

Gully erosion occurs when runoff water accumulates and rapidly flows in narrow channels during or immediately after heavy rains or melting snow, removing soil to a considerable depth.

Rivers and Streams

Dobbingstone Burn, Scotland—This photo illustrates two different types of erosion affecting the same place. Valley erosion is occurring due to the flow of the stream, and the boulders and stones (and much of the soil) that are lying on the edges are glacial till that was left behind as ice age glaciers flowed over the terrain.

Valley or *stream erosion* occurs with continued water flow along a linear feature. The erosion is both downward, deepening the valley, and headward, extending the valley into the hillside, creating head cuts and steep banks. In the earliest stage of stream erosion, the erosive activity is dominantly vertical, the valleys have a typical V cross-section and the stream gradient is relatively steep. When some base level is reached, the erosive activity switches to lateral erosion, which widens the valley floor and creates a narrow floodplain. The stream gradient becomes nearly flat, and lateral deposition of sediments becomes important as the stream meanders across the valley floor. In all stages of stream erosion, by far the most erosion occurs during times of flood, when more and faster-moving water is available to carry a larger sediment load. In such processes, it is not the water alone that erodes: suspended abrasive particles, pebbles and boulders can also act erosively as they traverse a surface, in a process known as *traction*.

Bank erosion is the wearing away of the banks of a stream or river. This is distinguished from changes on the bed of the watercourse, which is referred to as *scour*. Erosion and changes in the form of river banks may be measured by inserting metal rods into the bank and marking the position of the bank surface along the rods at different times.

Thermal erosion is the result of melting and weakening permafrost due to moving water. It can occur both along rivers and at the coast. Rapid river channel migration observed in the Lena River of Siberia is due to thermal erosion, as these portions of the banks are composed of permafrost-cemented non-cohesive materials. Much of this erosion occurs as the weakened banks fail in large slumps. Thermal erosion also affects the Arctic coast, where wave action and near-shore temperatures combine to undercut permafrost bluffs along the shoreline and cause them to fail. Annual erosion rates along a 100-kilometre (62-mile) segment of the Beaufort Sea shoreline averaged 5.6 metres (18 feet) per year from 1955 to 2002.

Floods

At extremely high flows, kolks, or vortices are formed by large volumes of rapidly rushing water. Kolks cause extreme local erosion, plucking bedrock and creating pothole-type geographical features called Rock-cut basins. Examples can be seen in the flood regions result from glacial Lake Missoula, which created the channeled scablands in the Columbia Basin region of eastern Washington.

Wind Erosion

Árbol de Piedra, a rock formation in the Altiplano, Bolivia sculpted by wind erosion.

Wind erosion is a major geomorphological force, especially in arid and semi-arid regions. It is also a major source of land degradation, evaporation, desertification, harmful airborne dust, and crop damage—especially after being increased far above natural rates by human activities such as deforestation, urbanization, and agriculture.

Wind erosion is of two primary varieties: *deflation*, where the wind picks up and carries away loose particles; and *abrasion*, where surfaces are worn down as they are struck by airborne particles carried by wind. Deflation is divided into three categories: (1) *surface creep*, where larger, heavier particles slide or roll along the ground; (2) *saltation*, where particles are lifted a short height into the air, and bounce and saltate across the surface of the soil; and (3) *suspension*, where very small and light particles are lifted into the air by the wind, and are often carried for long distances. Saltation is responsible for the majority (50–70%) of wind erosion, followed by suspension (30–40%), and then surface creep (5–25%). Silty soils tend to be the most affected by wind erosion; silt particles are relatively easily detached and carried away.

Wind erosion is much more severe in arid areas and during times of drought. For example, in the Great Plains, it is estimated that soil loss due to wind erosion can be as much as 6100 times greater in drought years than in wet years.

Mass Movement

Wadi in Makhtesh Ramon, Israel, showing gravity collapse erosion on its banks.

Mass movement is the downward and outward movement of rock and sediments on a sloped surface, mainly due to the force of gravity.

Mass movement is an important part of the erosional process, and is often the first stage in the breakdown and transport of weathered materials in mountainous areas. It moves material from higher elevations to lower elevations where other eroding agents such as streams and glaciers can then pick up the material and move it to even lower elevations. Mass-movement processes are always occurring continuously on all slopes; some mass-movement processes act very slowly; others occur very suddenly, often with disastrous results. Any perceptible down-slope movement of rock or sediment is often referred to in general terms as a landslide. However, landslides can be classified in a much more detailed way that reflects the mechanisms responsible for the movement and the velocity at which the movement occurs. One of the visible topographical manifestations of a very slow form of such activity is a scree slope.

Slumping happens on steep hillsides, occurring along distinct fracture zones, often within materials like clay that, once released, may move quite rapidly downhill. They will often show a spoon-shaped isostatic depression, in which the material has begun to slide downhill. In some cases, the slump is caused by water beneath the slope weakening it. In many cases it is simply the result of poor engineering along highways where it is a regular occurrence.

Surface creep is the slow movement of soil and rock debris by gravity which is usually not perceptible except through extended observation. However, the term can also describe the rolling of dislodged soil particles 0.5 to 1.0 mm (0.02 to 0.04 in) in diameter by wind along the soil surface.

Factors Affecting Soil Erosion

Climate

The amount and intensity of precipitation is the main climatic factor governing soil erosion by water. The relationship is particularly strong if heavy rainfall occurs at times when, or in locations where, the soil's surface is not well protected by vegetation. This might be during periods when agricultural activities leave the soil bare, or in semi-arid regions where vegetation is naturally sparse. Wind erosion requires strong winds, particularly during times of drought when vegetation is sparse and soil is dry (and so is more erodible). Other climatic factors such as average temperature and temperature range may also affect erosion, via their effects on vegetation and soil properties. In general, given similar vegetation and ecosystems, areas with more precipitation (especially high-intensity rainfall), more wind, or more storms are expected to have more erosion.

In some areas of the world (e.g. the mid-western USA), rainfall intensity is the primary determinant of erosivity, with higher intensity rainfall generally resulting in more soil erosion by water. The size and velocity of rain drops is also an important factor. Larger and higher-velocity rain drops have greater kinetic energy, and thus their impact will displace soil particles by larger distances than smaller, slower-moving rain drops.

In other regions of the world (e.g. western Europe), runoff and erosion result from relatively low intensities of stratiform rainfall falling onto previously saturated soil. In such situations, rainfall amount rather than intensity is the main factor determining the severity of soil erosion by water.

Soil Structure and Composition

Erosional gully in unconsolidated Dead Sea (Israel) sediments along the southwestern shore. This gully was excavated by floods from the Judean Mountains in less than a year.

The composition, moisture, and compaction of soil are all major factors in determining the erosivity of rainfall. Sediments containing more clay tend to be more resistant to erosion than those with sand or silt, because the clay helps bind soil particles together. Soil containing high levels of organic materials are often more resistant to erosion, because the organic materials coagulate soil colloids and create a stronger, more stable soil structure. The amount of water present in the soil before the precipitation also plays an important role, because it sets limits on the amount of water that can be absorbed by the soil (and hence prevented from flowing on the surface as erosive runoff). Wet, saturated soils will not be able to absorb as much rain water, leading to higher levels of surface runoff and thus higher erosivity for a given volume of rainfall. Soil compaction also affects the permeability of the soil to water, and hence the amount of water that flows away as runoff. More compacted soils will have a larger amount of surface runoff than less compacted soils.

Vegetative Cover

Vegetation acts as an interface between the atmosphere and the soil. It increases the permeability of the soil to rainwater, thus decreasing runoff. It shelters the soil from winds, which results in decreased wind erosion, as well as advantageous changes in microclimate. The roots of the plants bind the soil together, and interweave with other roots, forming a more solid mass that is less susceptible to both water and wind erosion. The removal of vegetation increases the rate of surface erosion.

Topography

The topography of the land determines the velocity at which surface runoff will flow, which in turn determines the erosivity of the runoff. Longer, steeper slopes (especially those without adequate vegetative cover) are more susceptible to very high rates of erosion during heavy rains than shorter, less steep slopes. Steeper terrain is also more prone to mudslides, landslides, and other forms of gravitational erosion processes.

Human Activities that Increase Soil Erosion

Agricultural Practices

Tilled farmland such as this is very susceptible to erosion from rainfall, due to the destruction of vegetative cover and the loosening of the soil during plowing.

Unsustainable agricultural practices are the single greatest contributor to the global increase in erosion rates. The tillage of agricultural lands, which breaks up soil into finer particles, is one of the primary factors. The problem has been exacerbated in modern times, due to mechanized agricultural equipment that allows for deep plowing, which severely increases the amount of soil that is available for transport by water erosion. Others include mono-cropping, farming on steep slopes, pesticide and chemical fertilizer usage (which kill organisms that bind soil together), row-cropping, and the use of surface irrigation. A complex overall situation with respect to defining nutrient losses from soils, could arise as a result of the size selective nature of soil erosion events. Loss of total phosphorus, for instance, in the finer eroded fraction is greater relative to the whole soil. Extrapolating this evidence to predict subsequent behaviour within receiving aquatic systems, the reason is that this more easily transported material may support a lower solution P concentration compared to coarser sized fractions. Tillage also increases wind erosion rates, by dehydrating the soil and breaking it up into smaller particles that can be picked up by the wind. Exacerbating this is the fact that most of the trees are generally removed from agricultural fields, allowing winds to have long, open runs to travel over at higher speeds. Heavy grazing reduces vegetative cover and causes severe soil compaction, both of which increase erosion rates.

Deforestation

In an undisturbed forest, the mineral soil is protected by a layer of *leaf litter* and an *humus* that cover the forest floor. These two layers form a protective mat over the soil that absorbs the impact of rain drops. They are porous and highly permeable to rainfall, and allow rainwater to slow percolate into the soil below, instead of flowing over the surface as runoff. The roots of the trees and plants hold together soil particles, preventing them from being washed away. The vegetative cover acts to reduce the velocity of the raindrops that strike the foliage and stems before hitting the ground, reducing their kinetic energy. However it is the forest floor, more than the canopy, that prevents surface erosion. The terminal velocity of rain drops is reached in about 8 metres (26 feet). Because forest canopies are usually higher than this, rain drops can often regain terminal velocity even after striking the canopy. However, the intact forest floor, with its layers of leaf litter and organic matter, is still able to absorb the impact of the rainfall.

In this clearcut, almost all of the vegetation has been stripped from surface of steep slopes, in an area with very heavy rains. Severe erosion occurs in cases such as this, causing stream sedimentation and the loss of nutrient rich topsoil.

Deforestation causes increased erosion rates due to exposure of mineral soil by removing the humus and litter layers from the soil surface, removing the vegetative cover that binds soil together, and causing heavy soil compaction from logging equipment. Once trees have been removed by fire or logging, infiltration rates become high and erosion low to the degree the forest floor remains intact. Severe fires can lead to significant further erosion if followed by heavy rainfall.

Globally one of the largest contributors to erosive soil loss in the year 2006 is the slash and burn treatment of tropical forests. In a number of regions of the earth, entire sectors of a country have been rendered unproductive. For example, on the Madagascar high central plateau, comprising approximately ten percent of that country's land area, virtually the entire landscape is sterile of vegetation, with gully erosive furrows typically in excess of 50 metres (160 ft) deep and 1 kilometre (0.6 miles) wide. Shifting cultivation is a farming system which sometimes incorporates the slash and burn method in some regions of the world. This degrades the soil and causes the soil to become less and less fertile.

Roads and Urbanization

Urbanization has major effects on erosion processes—first by denuding the land of vegetative cover, altering drainage patterns, and compacting the soil during construction; and next by covering the land in an impermeable layer of asphalt or concrete that increases the amount of surface runoff and increases surface wind speeds. Much of the sediment carried in runoff from urban areas (especially roads) is highly contaminated with fuel, oil, and other chemicals. This increased runoff, in addition to eroding and degrading the land that it flows over, also causes major disruption to surrounding watersheds by altering the volume and rate of water that flows through them, and filling them with chemically polluted sedimentation. The increased flow of water through local waterways also causes a large increase in the rate of bank erosion.

Climate Change

The warmer atmospheric temperatures observed over the past decades are expected to lead to a more vigorous hydrological cycle, including more extreme rainfall events. The rise in sea levels that has occurred as a result of climate change has also greatly increased coastal erosion rates.

Studies on soil erosion suggest that increased rainfall amounts and intensities will lead to greater rates of soil erosion. Thus, if rainfall amounts and intensities increase in many parts of the world as expected, erosion will also increase, unless amelioration measures are taken. Soil erosion rates are expected to change in response to changes in climate for a variety of reasons. The most direct is the change in the erosive power of rainfall. Other reasons include: a) changes in plant canopy caused by shifts in plant biomass production associated with moisture regime; b) changes in litter cover on the ground caused by changes in both plant residue decomposition rates driven by temperature and moisture dependent soil microbial activity as well as plant biomass production rates; c) changes in soil moisture due to shifting precipitation regimes and evapo-transpiration rates, which changes infiltration and runoff ratios; d) soil erodibility changes due to decrease in soil organic matter concentrations in soils that lead to a soil structure that is more susceptible to erosion and increased runoff due to increased soil surface sealing and crusting; e) a shift of winter precipitation from non-erosive snow to erosive rainfall due to increasing winter temperatures; f) melting of permafrost, which induces an erodible soil state from a previously non-erodible one; and g) shifts in land use made necessary to accommodate new climatic regimes.

Studies by Pruski and Nearing indicated that, other factors such as land use unconsidered, it is reasonable to expect approximately a 1.7% change in soil erosion for each 1% change in total precipitation under climate change.

Global Environmental Effects

Due to the severity of its ecological effects, and the scale on which it is occurring, erosion constitutes one of the most significant global environmental problems we face today.

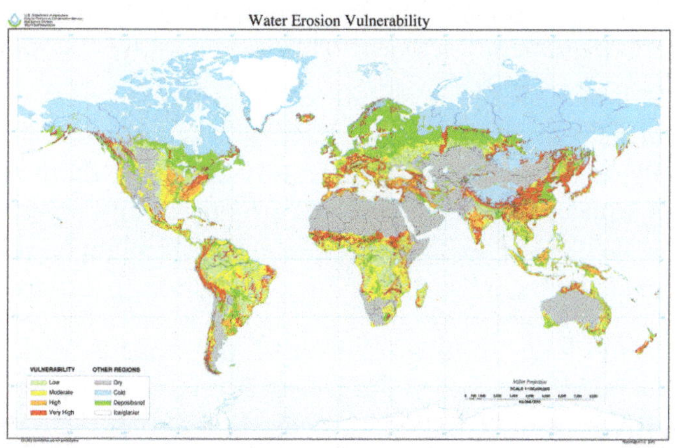

World map indicating areas that are vulnerable to high rates of water erosion.

During the 17th and 18th centuries, Easter Island experienced severe erosion due to deforestation and unsustainable agricultural practices. The resulting loss of topsoil ultimately led to ecological collapse, causing mass starvation and the complete disintegration of the Easter Island civilization.

Land Degradation

Water and wind erosion are now the two primary causes of land degradation; combined, they are responsible for 84% of degraded acreage.

Each year, about 75 billion tons of soil is eroded from the land—a rate that is about 13–40 times as fast as the natural rate of erosion. Approximately 40% of the world's agricultural land is seriously degraded. According to the United Nations, an area of fertile soil the size of Ukraine is lost every year because of drought, deforestation and climate change. In Africa, if current trends of soil degradation continue, the continent might be able to feed just 25% of its population by 2025, according to UNU's Ghana-based Institute for Natural Resources in Africa.

The loss of soil fertility due to erosion is further problematic because the response is often to apply chemical fertilizers, which leads to further water and soil pollution, rather than to allow the land to regenerate.

Sedimentation of Aquatic Ecosystems

Soil erosion (especially from agricultural activity) is considered to be the leading global cause of diffuse water pollution, due to the effects of the excess sediments flowing into the world's waterways. The sediments themselves act as pollutants, as well as being carriers for other pollutants, such as attached pesticide molecules or heavy metals.

The effect of increased sediments loads on aquatic ecosystems can be catastrophic. Silt can smother the spawning beds of fish, by filling in the space between gravel on the stream bed. It also reduces their food supply, and causes major respiratory issues for them as sediment enters their gills. The biodiversity of aquatic plant and algal life is reduced, and invertebrates are also unable to survive and reproduce. While the sedimentation event itself might be relatively short-lived, the ecological disruption caused by the mass die off often persists long into the future.

One of the most serious and long-running water erosion problems worldwide is in the People's Republic of China, on the middle reaches of the Yellow River and the upper reaches of the Yangtze River. From the Yellow River, over 1.6 billion tons of sediment flows into the ocean each year. The sediment originates primarily from water erosion in the Loess Plateau region of the northwest.

Airborne Dust Pollution

Soil particles picked up during wind erosion of soil are a major source of air pollution, in the form of airborne particulates—"dust". These airborne soil particles are often contaminated with toxic chemicals such as pesticides or petroleum fuels, posing ecological and public health hazards when they later land, or are inhaled/ingested.

Dust from erosion acts to suppress rainfall and changes the sky color from blue to white, which leads to an increase in red sunsets. Dust events have been linked to a decline in the health of coral reefs across the Caribbean and Florida, primarily since the 1970s. Similar dust plumes originate in the Gobi desert, which combined with pollutants, spread large distances downwind, or eastward, into North America.

Monitoring, Measuring and Modeling Soil Erosion

Terracing is an ancient technique that can significantly slow the rate of water erosion on cultivated slopes.

Monitoring and modeling of erosion processes can help people better understand the causes of soil erosion, make predictions of erosion under a range of possible conditions, and plan the implementation of preventative and restorative strategies for erosion. However, the complexity of erosion processes and the number of scientific disciplines that must be considered to understand and model them (e.g. climatology, hydrology, geology, soil science, agriculture, chemistry, physics, etc.) makes accurate modelling challenging. Erosion models are also non-linear, which makes them difficult to work with numerically, and makes it difficult or impossible to scale up to making predictions about large areas from data collected by sampling smaller plots.

The most commonly used model for predicting soil loss from water erosion is the *Universal Soil Loss Equation (USLE)*. This was developed in the 1960s and 1970s. It estimates the average annual soil loss A on a plot-sized area as:

$$A = RKLSCP$$

where R is the rainfall erosivity factor, K is the soil erodibility factor, L and S are topographic factors representing length and slope, C is the cover and management factor and P is the support practices factor.

Despite the USLE's plot-scale spatial focus, the model has often been used to estimate soil erosion on much larger areas, such as watersheds or even whole continents. For example, RUSLE has recently been used to quantify soil erosion across the whole of Europe . This is scientifically controversial, for several reasons. One major problem is that the USLE cannot simulate gully erosion, and so erosion from gullies is ignored in any USLE-based assessment of erosion. Yet erosion from gullies can be a substantial proportion (10–80%) of total erosion on cultivated and grazed land.

During the 50 years since the introduction of the USLE, many other soil erosion models have been developed. But because of the complexity of soil erosion and its constituent processes, all erosion models can give unsatisfactory results when validated i.e. when model predictions are compared with real-world measurements of erosion. Thus new soil erosion models continue to be developed. Some of these remain USLE-based, e.g. the G2 model . Other soil erosion models have largely (e.g. the Water Erosion Prediction Project model) or wholly (e.g. the Rangeland Hydrology and Erosion Model) abandoned usage of USLE elements.

Prevention and Remediation

A windbreak (the row of trees) planted next to an agricultural field, acting as a shield against strong winds. This reduces the effects of wind erosion, and provides many other benefits.

The most effective known method for erosion prevention is to increase vegetative cover on the land, which helps prevent both wind and water erosion. Terracing is an extremely effective means of erosion control, which has been practiced for thousands of years by people all over the world. Windbreaks (also called shelterbelts) are rows of trees and shrubs that are planted along the edges of agricultural fields, to shield the fields against winds. In addition to significantly reducing wind erosion, windbreaks provide many other benefits such as improved microclimates for crops (which are sheltered from the dehydrating and otherwise damaging effects of wind), habitat for beneficial bird species, carbon sequestration, and aesthetic improvements to the agricultural landscape. Traditional planting methods, such as mixed-cropping (instead of monocropping) and crop rotation have also been shown to significantly reduce erosion rates. Crop residues play a role in the mitigation of erosion, because they reduce the impact of raindrops breaking up the soil particles. There is a higher potential for erosion when producing potatoes than when growing cereals, or oilseed crops. Forages have a fibrous root system, which helps combat erosion by anchoring the plants to the top layer of the soil, and covering the entirety of the field, as it is a non-row crop. In tropical coastal systems, properties of mangroves have been examined as a potential means to reduce soil erosion. Their complex root structures are known to help reduce wave damage from storms and flood impacts while binding and building soils. These roots can slow down water flow, leading to the deposition of sediments and reduced erosion rates. However, in order to maintain sediment balance, adequate mangrove forest width needs to be present.

References

- Pruski, F. F.; Nearing, M. A. (2002). "Runoff and soil loss responses to changes in precipitation: a computer simulation study". Journal of Soil and Water Conservation. 57 (1): 7–16
- Poesen, Jean; et al. (2007). "Gully erosion in Europe". In Boardman, John; Poesen, Jean. Soil Erosion in Europe. John Wiley & Sons. pp. 516–519. ISBN 978-0-470-85911-7
- James K. B. Bishop; Russ E. Davis & Jeffrey T. Sherman (2002). "Robotic Observations of Dust Storm Enhancement of Carbon Biomass in the North Pacific". Science 298. pp. 817–821. Archived from the original on 2010-11-18. Retrieved 2009-06-20
- Science Daily (1999-07-14). "African Dust Called A Major Factor Affecting Southeast U.S. Air Quality". Retrieved 2007-06-10
- Toy, Terrence J.; et al. (2002). Soil Erosion: Processes, Prediction, Measurement, and Control. John Wiley & Sons. p. 1. ISBN 978-0-471-38369-7
- Curran, W. (2016). "Cover Crops for Conservation Tillage Methods". Penn State University. Retrieved December 1, 2016
- Stuart, Gordon W. & Edwards, Pamela J. (2006). "Concepts about forests and water". Northern Journal of Applied Forestry. 23 (1)
- Toy, Terrence J.; et al. (2002). Soil Erosion: Processes, Prediction, Measurement, and Control. John Wiley & Sons. pp. 60–61. ISBN 978-0-471-38369-7
- Cheraghi M., Jomaa S., Sander G. C., Barry D. A. (2016). "Hysteretic sediment fluxes in rainfall-driven soil erosion: Particle size effects". Water Resour. Res. 52. doi:10.1002/2016WR019314
- Nearing, M.A.; Norton, L.D.; Bulgakov, D.A.; Larionov, G.A.; West, L.T.; Dontsova, K.M. (1997). "Hydraulics and erosion in eroding rills". Water Resources Research. 33 (4): 865–876. Bibcode:1997WRR....33..865N. doi:10.1029/97wr00013
- Food and Agriculture Organization (1965). "Types of erosion damage". Soil Erosion by Water: Some Measures for Its Control on Cultivated Lands. United Nations. pp. 23–25. ISBN 978-92-5-100474-6
- Costard, F.; Dupeyrat, L.; Gautier, E.; Carey-Gailhardis, E. (2003). "Fluvial thermal erosion investigations along a rapidly eroding river bank: application to the Lena River (central Siberia)". Earth Surface Processes and Landforms. 28 (12): 1349–1359. Bibcode:2003ESPL...28.1349C. doi:10.1002/esp.592
- Jones, B.M.; Hinkel, K.M.; Arp, C.D.; Eisner, W.R. (2008). "Modern Erosion Rates and Loss of Coastal Features and Sites, Beaufort Sea Coastline, Alaska". Arctic. Arctic Institute of North America. 61 (4): 361–372. doi:10.14430/arctic44
- Poesen, Jean; et al. (2002). "Gully erosion in dryland environments". In Bull, Louise J.; Kirby, M.J. Dryland Rivers: Hydrology and Geomorphology of Semi-Arid Channels. John Wiley & Sons. ISBN 978-0-471-49123-1
- Evans, R (2012). "Assessment and monitoring of accelerated water erosion of cultivated land – when will reality be acknowledged?". Soil Use and Management. 29 (1): 105–118. doi:10.1111/sum.12010
- Panagos, P.; Ballabio, C.; Borrelli, P.; Meusburger, K.; Klik, A.; Rousseva, S.; Tadic, M.P.; Michaelides, S.; Hrabalíková, M.; Olsen, P.; Aalto, J.; Lakatos, M.; Rymszewicz, A.; Dumitrescu, A.; Beguería, S.; Alewell, C. (2015). "Rainfall Erosivity in Europe". Sci Total Environ. 511: 801–814. doi:10.1016/j.scitotenv.2015.01.008
- Borah, Deva K.; et al. (2008). "Watershed sediment yield". In Garcia, Marcelo H. Sedimentation Engineering: Processes, Measurements, Modeling, and Practice. ASCE Publishing. p. 828. ISBN 978-0-7844-0814-8
- Panagos, P.; Borrelli, P.; Meusburger (2015). "A New European Slope Length and Steepness Factor (LS-Factor) for Modeling Soil Erosion by Water". Geosciences, MDPI. pp. 117–126. doi:10.3390/geosciences5020117

Permissions

All chapters in this book are published with permission under the Creative Commons Attribution Share Alike License or equivalent. Every chapter published in this book has been scrutinized by our experts. Their significance has been extensively debated. The topics covered herein carry significant information for a comprehensive understanding. They may even be implemented as practical applications or may be referred to as a beginning point for further studies.

We would like to thank the editorial team for lending their expertise to make the book truly unique. They have played a crucial role in the development of this book. Without their invaluable contributions this book wouldn't have been possible. They have made vital efforts to compile up to date information on the varied aspects of this subject to make this book a valuable addition to the collection of many professionals and students.

This book was conceptualized with the vision of imparting up-to-date and integrated information in this field. To ensure the same, a matchless editorial board was set up. Every individual on the board went through rigorous rounds of assessment to prove their worth. After which they invested a large part of their time researching and compiling the most relevant data for our readers.

The editorial board has been involved in producing this book since its inception. They have spent rigorous hours researching and exploring the diverse topics which have resulted in the successful publishing of this book. They have passed on their knowledge of decades through this book. To expedite this challenging task, the publisher supported the team at every step. A small team of assistant editors was also appointed to further simplify the editing procedure and attain best results for the readers.

Apart from the editorial board, the designing team has also invested a significant amount of their time in understanding the subject and creating the most relevant covers. They scrutinized every image to scout for the most suitable representation of the subject and create an appropriate cover for the book.

The publishing team has been an ardent support to the editorial, designing and production team. Their endless efforts to recruit the best for this project, has resulted in the accomplishment of this book. They are a veteran in the field of academics and their pool of knowledge is as vast as their experience in printing. Their expertise and guidance has proved useful at every step. Their uncompromising quality standards have made this book an exceptional effort. Their encouragement from time to time has been an inspiration for everyone.

The publisher and the editorial board hope that this book will prove to be a valuable piece of knowledge for students, practitioners and scholars across the globe.

Index